Project Management beyond Waterfall and Agile

Best Practices and Advances in Program Management Series

Series Editor
Ginger Levin

RECENTLY PUBLISHED TITLES

Project Management beyond Waterfall and Agile
Mounir Ajam

Realizing Strategy through Projects: The Executive's Guide
Carl Marnewick

PMI-PBA® Exam Practice Test and Study Guide
Brian Williamson

Earned Benefit Program Management: Aligning, Realizing, and Sustaining Strategy
Crispin Piney

The Entrepreneurial Project Manager
Chris Cook

Leading and Motivating Global Teams: Integrating Offshore Centers and the Head Office
Vimal Kumar Khanna

Project and Program Turnaround
Thomas Pavelko

Project Portfolio Management in Theory and Practice: Thirty Case Studies from around the World
Jamal Moustafaev

Project Management in Extreme Situations: Lessons from Polar Expeditions, Military and Rescue Operations, and Wilderness Exploration
Monique Aubry and Pascal Lievre

Benefits Realization Management: Strategic Value from Portfolios, Programs, and Projects
Carlos Eduardo Martins Serra

IT Project Management: A Geek's Guide to Leadership
Byron A. Love

Situational Project Management: The Dynamics of Success and Failure
Oliver F. Lehmann

Project Management beyond Waterfall and Agile

Mounir A. Ajam

CRC Press
Taylor & Francis Group
Boca Raton London New York

CRC Press is an imprint of the
Taylor & Francis Group, an **Informa** business

AN AUERBACH BOOK

"AACE" is a registered trademark of the Association for the Advancement of Cost Engineers.
"GAPPS" is a registered trademark of the Global Alliance of Project Performance Standards.
"GPM" is a registered trademark and "P5" and "PRiSM" are trademarks of Green Project Management.
"IPMA," "ICB," and "OCB" are registered trademarks of the International Project Management Association.
"ISO" is a registered trademark of the International Organization for Standardization.
"PMBOK" and "PMP" are registered trademarks of the Project Management Institute.
"PRINCE2" is a registered trademark of AXELOS. All rights reserved.
"SAVE" is a registered trademark of the Society for Advancement of Value Engineers International.
"The SUKAD Way," "The Customizable and Adaptable Methodology for Managing Projects (CAMMP)," "The Seven Elements of Project Management Maturity (The 7Es)," "The Four Dimensions of Project Success," "The Four Control Reference Points," and "2SPI" are trademarks of SUKAD.

CRC Press
Taylor & Francis Group
6000 Broken Sound Parkway NW, Suite 300
Boca Raton, FL 33487-2742

© 2018 by Taylor & Francis Group, LLC
CRC Press is an imprint of Taylor & Francis Group, an Informa business

No claim to original U.S. Government works

Printed on acid-free paper

International Standard Book Number-13: 978-1-138-70563-0 (Hardback)

This book contains information obtained from authentic and highly regarded sources. Reasonable efforts have been made to publish reliable data and information, but the author and publisher cannot assume responsibility for the validity of all materials or the consequences of their use. The authors and publishers have attempted to trace the copyright holders of all material reproduced in this publication and apologize to copyright holders if permission to publish in this form has not been obtained. If any copyright material has not been acknowledged please write and let us know so we may rectify in any future reprint.

Except as permitted under U.S. Copyright Law, no part of this book may be reprinted, reproduced, transmitted, or utilized in any form by any electronic, mechanical, or other means, now known or hereafter invented, including photocopying, microfilming, and recording, or in any information storage or retrieval system, without written permission from the publishers.

For permission to photocopy or use material electronically from this work, please access www.copyright.com (http://www.copyright.com/) or contact the Copyright Clearance Center, Inc. (CCC), 222 Rosewood Drive, Danvers, MA 01923, 978-750-8400. CCC is a not-for-profit organization that provides licenses and registration for a variety of users. For organizations that have been granted a photocopy license by the CCC, a separate system of payment has been arranged.

Trademark Notice: Product or corporate names may be trademarks or registered trademarks, and are used only for identification and explanation without intent to infringe.

Visit the Taylor & Francis Web site at
http://www.taylorandfrancis.com

and the CRC Press Web site at
http://www.crcpress.com

Dedication

To Nabila, Sumer, and Akkad, my patient partners
along this exciting learning journey.

To project management practitioners everywhere
who strive for excellence!

Table of Contents

Dedication	*vii*
Contents	*ix*
List of Tables and Figures	*xxv*
Foreword	*xxvii*
Preface	*xxix*
Acknowledgments	*xxxiii*
About the Author	*xxxv*

Section I, Part A
The Global Project Management Scene — 1

Chapter 1   Definitions and Clarifications — 3
    1.1   Introduction — 3
    1.2   Standard — 3
    1.3   De Facto Standard — 4
    1.4   Framework — 4
    1.5   Method and Methodology — 4
    1.6   Is a Method (Methodology) Also a Standard? — 5
    1.7   System — 6
    1.8   PM versus OPM — 6
    1.9   The Guide — 6
    1.10 Closing Comments — 6

Chapter 2 Leading Global Associations — 7
- 2.1 Introduction — 7
- 2.2 Global Associations: General Project Management — 7
- 2.3 Global Associations: Specialties — 8
- 2.4 Other Relevant Topics — 8
- 2.5 Closing Comments — 9

Chapter 3 *PMBOK® Guide* Overview — 11
- 3.1 Introduction — 11
- 3.2 Brief History — 11
- 3.3 Relevant Highlights from the *PMBOK® Guide* — 12
  - 3.3.1 Process Groups
  - 3.3.2 Knowledge Areas — 13
  - 3.3.3 Generic Standard Document — 13
- 3.4 Myths about the *PMBOK® Guide* — 13
  - 3.4.1 The *PMBOK® Guide* Is About Best Practices — 13
  - 3.4.2 The *PMBOK® Guide* Is a Methodology — 14
- 3.5 Closing Comments — 15

Chapter 4 ISO 21500 Overview — 17
- 4.1 Introduction — 17
- 4.2 Development and Influence — 17
- 4.3 The Process Groups — 17
- 4.4 Subject Areas — 18
- 4.5 Processes — 19
- 4.6 ISO versus the *PMBOK® Guide* — 19
- 4.7 Closing Comments — 19

Chapter 5 Other Relevant PM Standards — 21
- 5.1 Introduction — 21
- 5.2 IPMA® — 21
- 5.3 GPM® Global — 22
- 5.4 Closing Part A — 22

Section I, Part B
The Challenges in Current Practices — 23

Chapter 6 Gaps and Opportunities — 25
- 6.1 Summary of Previous Chapters — 25
- 6.2 Transition, Understanding the Challenges — 25

6.3	Objectives	26
6.4	A New Perspective or Shifting Paradigms?	27

Chapter 7 Clearing a Critical Challenge — 29

7.1	Introduction	29
7.2	The Critical Challenge	29
	7.2.1 Process Groups and Project Life Cycle	29
	7.2.2 Opinion or Fact?	31
	7.2.3 So, Why the Confusion?	32
7.3	The Project Life Cycle	32
	7.3.1 Project Life Cycle Definition	32
	7.3.2 Other Life Cycles	33
	7.3.3 Project Life Cycle Is a Variable	33
	7.3.4 Whose Perspective?	34
7.4	Project Phases and Stages	34
7.5	Clearing the Confusion	34
	7.5.1 Phase Perspective	35
	7.5.2 Project Perspective	36
	7.5.3 Project and Stage Perspectives	38
	7.5.4 Can We Combine?	39
	7.5.5 Can We Consider a Project as a Program?	39
	7.5.6 How About Small-Simple Projects?	40
7.6	Closing Comments	40

Chapter 8 Various Challenges — 43

8.1	Introduction	43
8.2	What Is Missing or Not Covered Enough	43
	8.2.1 A Methodology	43
	8.2.2 Organizational System	44
	8.2.3 Tailoring and Customization	44
	8.2.4 Project Classification	44
	8.2.5 Templates and Forms	44
	8.2.6 Project Life Cycle	45
	8.2.7 Benefits Realization	45
8.3	Other Challenges in Published Guides	45
	8.3.1 Planning	46
	8.3.2 Project Change Management	46
	8.3.3 Project Success	46
	8.3.4 Pre-Project	47
8.4	Monitor and Control	48
8.5	Conclusion and Recommendations	48

Section II, Part C
Overview of the Offered Solution — 49

Chapter 9 Organizational Project Management — 51
- 9.1 What Should Organizations Use? — 51
- 9.2 OPM and Professional Associations — 51
- 9.3 What Are the Components of an OPMS? — 52
- 9.4 Closing Comments — 54

Chapter 10 The Three-Dimensional Model — 55
- 10.1 Introduction — 55
- 10.2 A Few Definitions — 55
  - 10.2.1 Universal — 55
  - 10.2.2 Method or Methodology — 56
  - 10.2.3 Project Type — 56
  - 10.2.4 Project Domain — 56
  - 10.2.5 Project Classification — 56
  - 10.2.6 Platform — 56
- 10.3 Principles of Building a Method(ology) — 57
- 10.4 The SUKAD CAMMP™ Model — 57
  - 10.4.1 The First Dimension — 58
  - 10.4.2 The Second Dimension — 58
  - 10.4.3 The Third Dimension — 58
- 10.5 Methodology Main Characteristics — 58
- 10.6 What Are the Key Features of CAMMP™? — 58
- 10.7 What Is Not New — 59
- 10.8 What Is New (or Not Well Known) — 60
- 10.9 Agile/Scrum/Waterfall — 61
- 10.10 Closing Comments — 62

Chapter 11 Rationale and Model Perspectives — 63
- 11.1 The Story That Led to This Work — 63
- 11.2 The Challenges for Professionals — 64
- 11.3 The Birth of CAMMP™ — 64
- 11.4 Full Project Life Cycle Perspective — 64
- 11.5 Multiple Stakeholders — 65
- 11.6 Idea Developer (Project Owner) Perspective — 65
- 11.7 Closing Comments — 65

Chapter 12 A Vital Concept: The Stage Gate Process — 67
- 12.1 Project Management Is a Disciplined Approach — 67
- 12.2 Definition of Gates — 68

12.3	Project Life Cycle and Stage Gates	68
12.4	Importance of the Stage Gate Process	68
12.5	Governance	69
12.6	Is Not the Stage Gate Process . . .	70
	12.6.1 . . . A Form of Risk Aversion?	70
	12.6.2 . . . A Form of Bureaucratic Process?	70
	12.6.3 . . . A Killer of Innovation?	70
	12.6.4 Case Study	71
12.7	Closing Comments	72

Chapter 13 Brief Overview of CAMMP™ — 73

13.1	Introduction	73
13.2	History of CAMMP™	73
13.3	The Model's Brief Explanation	74
	13.3.1 Overview	74
	13.3.2 Phases	74
	13.3.3 Stages	74
	13.3.4 Stage Deliverables and Gates	75
13.4	Summary of Phases and Stages	76
	13.4.1 Discovery Phase	76
	13.4.2 A Message on Authorization	77
	13.4.3 Development Phase	77
	13.4.4 Delivery Phase	78
13.5	The Full Standard Model	79
13.6	The Strategic Aspects in CAMMP™	80
13.7	Closing Comments	81

Section II, Part D
The Second Dimension (Project Management Processes) — 83

Chapter 14 Links to Global Standards — 85

14.1	Introduction to Part D	85
14.2	CAMMP™ Refresher	85
14.3	Professional Associations and Gaps	86
14.4	CAMMP™ Link to PMI and ISO	87
14.5	CAMMP™ Link to IPMA and Other Associations	87
14.6	How CAMMP™ Differs from Global References	87
	14.6.1 General Statements	87
	14.6.2 Terminology Differences	88
	14.6.3 Additional or Emphasized Concepts	88
	14.6.4 Shift in Planning and Additional Processes	89
14.7	Closing Comments	90

Chapter 15 Management and Detailed Planning   91
15.1 Introduction   91
15.2 One or More Plans?   92
15.3 Splitting Planning   92
15.4 Closing Comments   92

Chapter 16 Processes and Functions   93
16.1 Introduction   93
16.2 Project Management Processes   94
    16.2.1 The CAMMP™ Processes   94
    16.2.2 Process Gates   95
    16.2.3 Typical Process Within a Stage   95
16.3 Project Management Functions   95
    16.3.1 CAMMP™ Perspective   95
16.4 Closing Comments   97

Chapter 17 Processes and Deliverables Within a Stage   99
17.1 Overview   99
17.2 Authorize Process   100
    17.2.1 Develop Stage Authorization Document   100
    17.2.2 Establish Stage Success Criteria   100
    17.2.3 Mobilize Project Manager   100
    17.2.4 Identify Stakeholders   101
17.3 Plan Management Process   101
17.4 Plan Details Process   101
17.5 Implement Process   101
17.6 Control Process   103
17.7 Close Process   103
17.8 Closing Comments   103

Section II, Part E
The First Dimension (Project Life Cycle – Details)   105

Chapter 18 Discovery Phase – Concept Stage   107
18.1 Phase Overview   107
18.2 Stage Overview   107
18.3 Sequence of Events   108
18.4 Special Note About Processes   108
18.5 The Project Brief   109
    18.5.1 Overview   109
    18.5.2 Potential Pitfall   109
    18.5.3 Content   110

18.6	Stage Gate 1 (SG1)	110
18.7	Stage Summary and Next Steps	111

Chapter 19 Discovery Phase – Feasibility Stage — 113

19.1	Stage Overview	113
19.2	Sequence of Events	113
19.3	The Feasibility Study	114
	19.3.1 Introduction	114
	19.3.2 Feasibility Study and Risk Management	115
	19.3.3 The Content	115
	19.3.4 Feasibility Study Potential Pitfall	119
19.4	Stage Gate Two (SG2)	120
19.5	Project Authorization Document	120
	19.5.1 Purpose of the Project Authorization Document	120
	19.5.2 Content of the Project Authorization Document	121
19.6	Stage Summary	122

Chapter 20 Development Phase – Requirements Stage — 123

20.1	Phase Overview	123
20.2	Stage Overview	123
20.3	Sequence of Events	123
20.4	Project Requirements Document	124
	20.4.1 Overview	124
	20.4.2 Why the PRD?	124
	20.4.3 PRD Content	125
20.5	Stage Gate 3 (SG3)	129
	20.5.1 Overview	129
	20.5.2 Importance of This Stage Gate	129
20.6	Stage Summary	129

Chapter 21 Development Phase – Strategy Stage — 131

21.1	Stage Overview	131
21.2	Sequence of Events	131
21.3	Project Planning	132
21.4	Project Management Plan	132
	21.4.1 Sections of the Project Management Plan	132
	21.4.2 The Strategy Section	133
	21.4.3 The "How-To" Section	134
	21.4.4 People Aspects	137
	21.4.5 Class 2 Estimate Section	140
21.5	Once Again: Process Groups	140
21.6	Stage Gate 4 (SG4)	141

21.7	Advance Funding	141
	21.7.1 How Was the Project Funded up to This Point?	141
	21.7.2 Why No Final Funding Now?	142
21.8	Stage Summary	142

Chapter 22 Development Phase – Definition Stage — 143

22.1	Stage Overview	143
22.2	Sequence of Events	144
22.3	Project Detailed Plan	144
	22.3.1 Introduction	144
	22.3.2 Special Conditions	145
	22.3.3 Scope of Work	145
	22.3.4 Q/HSE	146
	22.3.5 Schedule	147
	22.3.6 Cost Estimating	148
	22.3.7 Staffing	150
	22.3.8 Risk Management	150
	22.3.9 Sustainability	150
	22.3.10 Project Procurement	150
	22.3.11 Other Considerations for the Project Detailed Plan	151
22.4	Request for Final Approval	151
22.5	Stage Gate 5 (SG5)	151
22.6	Stage Summary	152

Chapter 23 Delivery Phase – Implementation Stage — 153

23.1	Phase Overview	153
23.2	Stage Overview	154
23.3	Stage Sequence	154
23.4	Implementation and Control	154
	23.4.1 Overview	154
	23.4.2 Performance Management	155
	23.4.3 Start Implementation	157
	23.4.4 The Trip (A Simple Project)	157
23.5	Stage vs. Process Group	160
23.6	Operational Readiness	161
23.7	Stage Gate 6 (SG6)	161
23.8	Stage Gate 7 (SG7)	161
23.9	Stage Summary	162

Chapter 24 Delivery Phase – Operational Readiness Stage — 163

24.1	Stage Overview	163
24.2	Special Considerations	164

24.3	Stage Sequence	164
24.4	Pre-Handover Activities	164
24.5	Handover Activities	165
24.6	Stage Gate 6 (SG6)	165
24.7	Stage Summary	166

Chapter 25 Delivery Phase – Initial Operations Stage 167

25.1	Stage Overview	167
25.2	Stage Sequence	167
25.3	Pilot	168
	25.3.1 Example 1: A New Technology	168
	25.3.2 Example 2: A Hotel Project	168
	25.3.3 Example 3: An Organizational Change Project	168
25.4	Initial Operations	169
25.5	Stage Gate 8 (SG8)	169
25.6	Stage Summary	169

Chapter 26 Delivery Phase – Close Stage 171

26.1	Stage Overview	171
26.2	Stage Sequence	171
26.3	Stage vs. Process Group	171
26.4	Importance of Proper Project Closure	172
26.5	The Close-Out Report	173
	26.5.1 Reconciliation	173
	26.5.2 Lessons Learned	174
	26.5.3 Project Success Assessment	175
	26.5.4 Benefits Realization Evaluation	175
	26.5.5 Organizational Records Update	176
26.6	Stage Gate 9 (SG9)	176
26.7	Stage Summary	177

Section II, Part F
The Third Dimension (Advanced Topics) 179

Chapter 27 Introduction to the Third Dimension 181

27.1	A Refresher	181
27.2	The Third Dimension	182
27.3	What Are These Layers?	182

Chapter 28 Project Success 183

28.1	Introduction	183
28.2	Gaps in Practice	183

28.3 Perspectives — 183
      28.3.1 Strategic and Organizational Perspective — 183
      28.3.2 Project Owner Perspective — 184
   28.4 CAMMP™ Proposed Approach — 184
      28.4.1 Product Delivery Success: Technical — 184
      28.4.2 Project Management Success — 185
      28.4.3 Special Notices — 186
      28.4.4 Project Delivery Success — 186
      28.4.5 Objectives Success — 187
   28.5 Applying the Concept — 189
      28.5.1 Case Study — 189
      28.5.2 Project Brief and Project Success — 189
   28.6 When to Define the Criteria for Success — 190
   28.7 Closing Comments — 190

Chapter 29 Competence — 193
   29.1 Overview — 193
   29.2 Resources — 193
   29.3 IPMA ICB — 194
   29.4 Common Misunderstanding — 194
   29.5 Competence in Organizations — 195
      29.5.1 The Current State of Practice — 195
      29.5.2 Implementing Competence Development — 195
   29.6 Competence and CAMMP™ — 196

Chapter 30 Sustainability — 197
   30.1 Special Message — 197
   30.2 Overview — 197
   30.3 Resources — 197
   30.4 P5 — 198
   30.5 PRiSM™ — 198
   30.6 Sustainability and CAMMP™ — 198

Chapter 31 Best Practices — 201
   31.1 Overview — 201
   31.2 Resources — 201
   31.3 The Value of Best Practices — 202
   31.4 Best Practices and CAMMP™ — 202

Section III, Part G
Practical Management of the Three Dimensions — 203

Chapter 32 Managing Across the Stages — 205

32.1	Introduction to Section III	205
32.2	Reflections on Challenges	206
32.3	The Charter versus Managing Across the Stages	207
32.4	Simulating Managing Across the Stages	208
	32.4.1 Overview	208
	32.4.2 General Explanation	208
	32.4.3 Summary of the Project	212
32.5	A Project Example, End-to-End	213
	32.5.1 The Project	213
	32.5.2 Discovery Phase	213
	32.5.3 Reflections	214
	32.5.4 Development Phase – Requirements Stage	214
	32.5.5 Summary of the Rest of the Stages	215
	32.5.6 General Comments	216
32.6	Closing Comments	216

Chapter 33 Project Approvals — 217

33.1	Introduction	217
33.2	Types of Approvals	217
	33.2.1 Stage Gate Approvals	217
	33.2.2 Process Gates Approvals	218
	33.2.3 Budget Approvals	218
	33.2.4 Timing of Approvals	219
33.3	Budgeting Common Practices	220
	33.3.1 Preset Budget	220
	33.3.2 Project Budget (Internal and External Costs)	220
	33.3.3 Constrained Budget	220
33.4	Leading Practices	220
	33.4.1 Is There a Leading Practice?	220
	33.4.2 What Is the Leading Practice?	220
	33.4.3 Stage Gates versus Budget Approvals	221
33.5	Three Budget Approvals	221
	33.5.1 First Budget Approval	221
	33.5.2 Second Budget Approval	222
	33.5.3 Third Budget Approval	223
33.6	Closing Comments	223

Chapter 34 Project Estimates — 225

34.1	Introduction	225
34.2	Stage versus Project Estimates	225
34.3	Proposed Approach	226
	34.3.1 Overview	226

34.3.2 Class 1 Estimate	226
34.3.3 Class 2 Estimate	227
34.3.4 Class 3 Estimate	227
34.3.5 Summary of the Three Estimates	228
34.4 Are Three Estimates Too Many?	228
34.5 Components of a Good Estimate	228
34.6 Estimating and Project Management Maturity	229
34.7 Estimating and the Project Life cycle	230
34.8 Closing Comments	231

Chapter 35 Project Control — 233

35.1 Introduction	233
35.2 Baseline	233
35.3 Stage versus Project Control	233
35.3.1 Stage Control	233
35.3.2 Project Control	234
35.4 CAMMP™ Alternative Perspective	235
35.4.1 Overview	235
35.4.2 First Control Reference Point	235
35.4.3 Second Control Reference Point	236
35.4.4 Third Control Reference Point	237
35.4.5 Fourth Control Reference Point	237
35.4.6 Why Four Points?	237
35.5 Project Management Maturity and Control	238

Chapter 36 Project Change Management — 239

36.1 Introduction	239
36.2 Agile/Scrum	239
36.3 Why Change Management Is Critical	240
36.4 The Hidden Changes – Scope Creep	240
36.4.1 Overview	240
36.4.2 Discovery Phase	241
36.4.3 Requirements/Strategy Stages	241
36.4.4 Definition Stage	241
36.4.5 Why Scope Creep Happens	242
36.5 Change Management in the Fog	242
36.6 Traditional Change Management	243
36.7 Types of Changes	243
36.8 Closing Comments	244

Chapter 37 Project Risk Management — 245

37.1 Introduction	245
37.2 Threats or Threats and Opportunities?	245

37.3	Frequency of Performing Risk Assessment	245
37.4	Stage versus Project Risks	246
37.5	Risk Management Across the Stages	246
	37.5.1 Discovery Phase	246
	37.5.2 Other Project Stages	248
	37.5.3 Timing for Managing Risks	248

Chapter 38 The People Aspects — 249

38.1	The Project Stakeholders	249
38.2	Stakeholders Across the Project Stages	249
38.3	The Bus Trip Analogy	250
38.4	The Project Team	251

Chapter 39 All Other Topics and Layers — 253

Section III, Part H
Practical Real-World Application — 255

Chapter 40 Introduction to Applying CAMMP™ — 257

40.1	Introduction	257
40.2	Specialization	257
40.3	Project Samples	258
	40.3.1 Sample 1	258
	40.3.2 Sample 2	258
	40.3.3 Sample 3	258
	40.3.4 Sample 4	259
40.4	Closing Comments	259

Chapter 41 Managing per the Project's Class — 261

41.1	Project Classification	261
41.2	How to Measure Size and Complexity	262
	41.2.1 Size	262
	41.2.2 Complexity	262
	41.2.3 Summary	262
	41.2.4 Scoring Model	262
	41.2.5 Special Situations	263
41.3	Classification in Project Management Guides	263
41.4	How to Manage per a Given Classification	264
	41.4.1 Overview	264
	41.4.2 Small-Simple Projects	265
	41.4.3 Medium-Moderate Projects	265
	41.4.4 Large-Complex Projects	266
	41.4.5 Blended Approach	266
41.5	Closing Comments	267

Chapter 42 Life Cycles 360° — 269

- 42.1 Introduction — 269
- 42.2 Product Life Cycle — 269
  - 42.2.1 Defining "Product" — 269
  - 42.2.2 Product Life Cycle — 270
  - 42.2.3 Product Life Cycle: Textbook Perspective — 270
- 42.3 Project Life Cycle — 271
- 42.4 Service Provider Perspective — 271
  - 42.4.1 Overview — 271
  - 42.4.2 The Sample Project — 272
  - 42.4.3 How Many Stages? — 272
  - 42.4.4 Discovery Phase — 272
  - 42.4.5 Requirements Stage — 273
  - 42.4.6 Strategy Stage — 273
  - 42.4.7 Definition Stage — 273
  - 42.4.8 Implementation Stage — 274
  - 42.4.9 Operational Readiness Stage — 274
  - 42.4.10 Close Stage — 274
- 42.5 General Observations — 274
  - 42.5.1 CAMMP™ Works for Both Perspectives
  - 42.5.2 Two Projects or Two Phases? — 275
- 42.6 Closing Comments — 275

Chapter 43 Customizing and Adapting CAMMP™ — 277

- 43.1 Need to Tailor CAMMP™ — 277
- 43.2 Example to Differentiate — 277
- 43.3 Why Customize and Adapt the Model — 278
- 43.4 What Is Customizable and Adaptable? — 279
- 43.5 Customizing to a Project Domain (Industry) — 279
- 43.6 Common Concepts — 279
- 43.7 How to Customize — 280
  - 43.7.1 Similar Projects — 280
  - 43.7.2 Different Projects — 280
  - 43.7.3 Who Will Customize — 281
  - 43.7.4 A Reviewer Comment — 281
- 43.8 Project Manager Authority — 281
- 43.9 A Story Reflecting Current Reality — 282
- 43.10 Potential Pitfalls — 282
- 43.11 Closing Comments — 283

Chapter 44 Simulating CAMMP™ for Various Projects — 285
- 44.1 Introduction — 285
- 44.2 Capital Investment Project — 285
  - 44.2.1 The Project Life Cycle — 285
  - 44.2.2 Discovery Phase — 286
  - 44.2.3 Development Phase – Part 1 — 286
  - 44.2.4 Development Phase – Part 2 — 286
  - 44.2.5 Implementation Stage — 287
  - 44.2.6 Operational Readiness Stage — 287
  - 44.2.7 Initial Operations Stage — 288
  - 44.2.8 Close Stage — 288
  - 44.2.9 Project Success — 288
- 44.3 Writing and Publishing a Book — 289
  - 44.3.1 Discovery Phase — 289
  - 44.3.2 Requirements Stage — 289
  - 44.3.3 Strategy Stage — 289
  - 44.3.4 Definition Stage — 289
  - 44.3.5 Implementation Stage — 289
  - 44.3.6 Operational Readiness Stage — 290
  - 44.3.7 Project Close Stage — 290
  - 44.3.8 Project Success — 290
- 44.4 Closing Comments — 291

Chapter 45 Closing Chapter — 293
- 45.1 Introduction — 293
- 45.2 The Author and Basis for the Book — 293
- 45.3 Objectives — 294
- 45.4 A New Perspective or Shifting Paradigms? — 295
- 45.5 The Bottom Line — 295
- 45.6 The Future — 296
  - 45.6.1 Asset Management — 296
  - 45.6.2 Sample Projects — 296
  - 45.6.3 OPMS — 296
- 45.7 Author's Final Statement — 297

Appendix A: The CAMMP™ Phases — 299
Appendix B: The CAMMP™ Stages — 301
Appendix C: The CAMMP™ Stage Deliverables — 303
Appendix D: The CAMMP™ Stage Gates — 307

Appendix E: The CAMMP™ Processes   309

Appendix F: The CAMMP™ Process Gates   311

Appendix G: Frequently Asked Questions   313

Acronyms   319

Glossary of Terms   321

Bibliography   325

Index   327

List of Tables and Figures

Tables

Table 3.1	Summary of Some of *The PMBOK® Guide* Changes	12
Table 17.1	Stage Processes, Deliverables, and Gates	99
Table 17.2	CAMMP™ Processes Step by Step	102
Table 41.1	Project Classification Summary, Categories, and Factors	263

Figures

Figure 4.1	The PDCA Cycle (Plan-Do-Check-Act)	18
Figure 4.2	The PDCA Cycle with Initiate and Close	18
Figure 7.1	How some practitioners of project management understand the project life cycle	30
Figure 7.2	The process groups within a typical project phase/stage	35
Figure 7.3	A project life cycle, with the initiating process group	36
Figure 7.4	A project life cycle, with the planning process group	36
Figure 7.5	A project life cycle, with the implementing process group	37
Figure 7.6	A project life cycle, adding the controlling process group	37
Figure 7.7	The project life cycle and repeating process groups	38
Figure 7.8	Mapping the process groups to a generic project life cycle	41
Figure 9.1	The Seven Elements of Project Management Maturity™	53
Figure 10.1	The Customizable and Adaptable Methodology for Managing Projects™ (CAMMP™)	59

Figure 10.2	Possible project life cycle model integrating adaptive principles	62
Figure 13.1	CAMMP™ project life cycle, with the three phases	74
Figure 13.2	CAMMP™ project life cycle, with phases and stages (per the standard model)	75
Figure 13.3	CAMMP™ project life cycle, with phases, stages, and stage gates	79
Figure 16.1	The six processes per the CAMMP™ Model	94
Figure 16.2	A typical process flowchart for a given stage, the main processes	96
Figure 18.1	Discovery Phase, process overview	108
Figure 19.1	Content of the feasibility study	116
Figure 20.1	Content of the project requirements document	125
Figure 20.2	High-level PBS (house example)	127
Figure 21.1	The project management plan with sections and subsections	133
Figure 21.2	Project management team: a typical structure for capital projects	138
Figure 21.3	The extended project team	139
Figure 23.1	The Delivery Phase, per CAMMP™ Standard Model	153
Figure 23.2	A sample time-phased budget and performance management baseline	156
Figure 28.1	Project success, first dimension: product delivery success	185
Figure 28.2	Project success, second dimension: project management success	185
Figure 28.3	Project success, third dimension: project delivery success	186
Figure 28.4	Project success, fourth dimension: objectives success	188
Figure 32.1	The project charter and stage charters	207
Figure 32.2	Sample large-complex project, Discovery Phase	209
Figure 32.3	Sample large-complex project, Development Phase	210
Figure 32.4	Sample large-complex project, Delivery Phase	211
Figure 33.1	Budget approvals across the project life cycle	222
Figure 34.1	Project estimates across the project life cycle	226
Figure 34.2	Estimate classification and level of scope definition	230
Figure 35.1	Stage control and stage control reference points	234
Figure 35.2	Project control with the concept of moving baseline (control reference points)	235
Figure 41.1	The CAMMP™ model for small-simple projects	264
Figure 41.2	The CAMMP™ model for medium-moderate projects	265
Figure 41.3	The CAMMP™ model for large-complex projects	266
Figure 42.1	Project life cycle, service provider perspective	275
Figure 44.1	Sample project life cycle for a capital project	286

Foreword

Projects are synonymous with advancement. The growth and development of human advancement throughout the ages would not have been possible without them.

Leading organizations understand the importance of projects and even more so how to effectively manage them. In the same way that upstarts and disruptors can change an industry, a well-managed project can be the catalyst for organizational success well into the future.

As products and services evolve with the times to be more efficient, faster, and sustainable, approaches to project management have largely remained the same.

I had the great pleasure to collaborate with SUKAD founder Mounir Ajam in 2014. His insights into the diverse approaches and standards that are used around the world and the work that was put into The SUKAD Way™ was incredible. SUKAD pushes the envelope on learning, development, frameworks, and models to ensure that the management of projects is sensible and most importantly leads to positive results.

In this latest book, Mounir has dived even deeper into organizational project management systems with the latest and more comprehensive CAMMP™ Model. I have always been of the mind that a project manager should be versed in many models so that the approach can be custom fit to the project, and CAMMP™ is one that should be a staple.

This book goes beyond the paint by numbers approach, transcending the "how" of project management to the "what" and "why," which is critical for leaders of change.

— Dr. Joel B. Carboni
President and Founder, GPM Global
President, IPMA-USA

Preface

Over the last few decades, the field of project management has seen significant growth and acceptance across sectors, industries, and domains. Project management skills are in demand across organizational levels. There are numerous certifications in project management—maybe more than fifty—offered by various professional associations and private companies. It would be safe to say that there are more than a few million practitioners with one project management certificate or another. Furthermore, in 2016 we witnessed the government of the United Kingdom recognizing project management by granting Chartered Status to the Association of Project Management; the United States also passed a law for projects and programs accountability within the US Government. Furthermore, there is PM^2, a project management methodology developed by and for the European Commission, which includes certifications for their staff.

Nevertheless, projects continue to fail or face many challenges.

Many executives still have not accepted or respected project management enough to make it a formal and permanent function in their organizations. For example, positions such as chief project officer or head of projects are still exceedingly rare. Also, project management departments, outside the capital projects industries, are a rare phenomenon. There are project management offices (PMOs), but many of these fail within two or three years from inception or do not deliver outstanding value to organizations. However, it is important to state that some organizations are becoming better in building their project management offices.

It is still common to observe the accidental project manager's syndrome. An accidental project manager is a professional who stumbles into project management, leaving behind project outcomes that are less than optimal. In other words, in these organizations, anyone can manage projects, but when it comes to finance, strategy, engineering, human resources, or marketing, these organizations insist on hiring or training specialists with a background in these domains.

Furthermore, we observe organizations that might invest in training their staff on project management topics. Management even encourages or mandates their employees to achieve one certification or another. However, they do not invest into building proper, comprehensive, and sustainable project management systems for their organization's various functions. Some of these organizations think that a 2, 3, or 5-day class would be enough to "master project management."

It is also unfortunate that some professional associations have pushed their certifications to the level of becoming a commodity, where young and inexperienced professionals can follow a jam-packed study program and become certified without having managed a project either in their career or in their personal lives. It is also not uncommon for some of these individuals to memorize a standard or study guide without knowing how to apply the concepts on a real, even simple project.

"Projects fail, blame the methodology."

"Projects are challenged, blame an incompetent vendor or supplier."

"Projects do not achieve their objectives; it must due to be the unforeseen conditions, the unknown-unknowns."

Individuals and organizations do not realize that this current state of practice is not sustainable. They do not fully understand that the root causes include lack of standardized processes, missing methods or methodological approach, and no real organizational system for managing projects. The unfortunate truth is that global project management associations did not even see the necessity to have guides or standards to build organizational project management systems until two or three years ago, and even when publishing such standards, they do not promote them heavily, perhaps because they are not revenue generators.

In this book, we will introduce such a system, but the core focus will be on some of the elements of such a system—a practical and systematic methodological approach for managing and delivering all types of projects. The methodological approach is *the Offered Solution*. In a future book, we will elaborate on and offer a model for building a comprehensive and sustainable organizational project management system (OPMS).

This book presents a three-dimensional delivery model that integrates the best learning from the various global associations in the field as understood and applied by the author. We have worked to deliver to you a book by a practitioner with more than three decades of project management experience working around the globe, in different roles, on various types of projects, of all sizes and complexities.

The three-dimensional delivery model consists of a project life cycle approach as the first dimension. The first dimension is a must for managing a project, regardless of size and complexity. The model also incorporates project management processes and functions as the second dimension. Together, the first two dimensions can lead to good or very good performance. However, excellence requires blending in a few layers of advanced topics such as best practices, project success, competence, and sustainability.

Although the focus of this book is a project management approach, it is vital to stress that the offered solution (The Customizable and Adaptable Methodology for Managing Projects™, CAMMP™) is a product (or service) delivery *methodology* and the processes and approach covered in this book are not limited to project management processes.

This work is a practical book, by a practitioner, to practitioners!

Structure of the Book

This book's core focus is to introduce a three-dimensional model for managing projects effectively. However, before presenting the model—the *offered solution*—the book addresses the *current reality* of the project management practice, including gaps, inconsistencies, and even errors. After the first section, the second section provides the reader with an overview of the offered solution and the *details of each of the three dimensions* of the methodological approach. The third section shifts to *applying the solution* by offering samples and suggestions on the necessity of tailoring the approach to best fit the organizational context.

The following image presents the high-level structure of the book, showing the three sections and eight parts. Each part consists of three or more chapters; refer to the table of contents for the detailed outline.

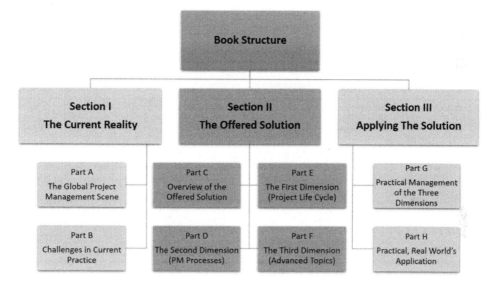

The first section represents the current reality of the project management practice. It consists of two parts:

- Part A presents the *global project management scene*, with a focus on project management professional associations and various definitions that are necessary for the rest of the book.
- Part B highlights the *challenges in the current practice* of project management, especially as they relate to the leading standards documents. This part clarifies some gaps, opportunities, and confusion between process groups and project phases versus the project life cycle. It also covers various challenges and misconceptions.

The second section presents *the offered solution*—a methodological approach—and is the core section of the book. It consists of four parts.

- Part C presents an *overview of the offered solution* and its rationale. The part includes topics such as the necessity to have an organizational project management system (OPMS) and introduces the three-dimensional model, its fundamental concepts, and the rationale behind it. The presented model is The Customizable and Adaptable Methodology for Managing Projects™ (CAMMP™).

- Part D highlights the *second dimension: project management processes and functions*. This part includes the links to global standards and guides, along with a new perspective on project planning and the need to split management planning from detailed (product-oriented) planning. Furthermore, this part includes two chapters dedicated to content not covered in the PMI and ISO guides, such as two additional subject areas and numerous other processes. These are essential topics, which will enable the reader to understand the guides and CAMMP™ fully and without any ambiguity.
- Part E shifts the focus and goes into the details of *the first dimension: project life cycle*. This dimension is about the concept of a project life cycle, spanning a project from the idea to closure. It includes a discussion of the phases, stages, stage deliverables, and stage gates. There is a dedicated chapter for each of the project stages of the standard CAMMP™ model.
- Part F addresses *the third dimension: advanced topics;* it includes topics that should overlay the first two dimensions. These are the subjects that can help organizations reach a higher level of project management maturity. These chapters cover competence, sustainability, project success, and best practices.

The third and last section of the book covers discussions for applying the methodological approach in the real world and under different conditions; it consists of two parts.

- Part G is titled *Practical Management of the Three Dimensions*; it elaborates and offers topics that help practitioners visualize the link between the dimensions and how to apply them. These include the following concepts: approvals, estimating, control, change and risk management, and the people aspects (teams and stakeholders).
- Part H, *Practical, Real-World Applications,* is about tailoring the CAMMP™ concepts to real-world scenarios. These chapters cover project classifications, project owner versus service provider perspectives, tailoring the model (customizing it to industries or organizations and adapting it to functions or project types), potential pitfalls, and two simulated projects. This part also includes the book's closing chapter.

The various appendices can be used as a quick reference guide to the book; these appendices include briefs on the project phases, stages, stage gates, process gates, and other relevant information.

— Mounir A. Ajam
1 September 2017

Acknowledgments

General Acknowledgments

The indirect influence of the project management model that I present in this writing project goes back to my early career at Exxon Chemical. I credit my first team leader, Mr. Ralph Spears. Ralph was a good friend and mentor. In addition, I recognize Mr. Ed Boyle, Mr. Jack Foerst, Mr. Charlie Thompson, Mr. Tony Leyesa, Mr. Jimmy McGregor, and Mr. Tom Sinkovic. It was during that early stage of my career that I had learned the value of a *disciplined approach to project management*.

Acknowledgment for the Methodology Development

Version 1

My colleagues Luc Bauwmans and Nada Chaban had a direct impact on enhancing the first version of the project management methodology, which I describe in this book.

Other contributors to the model are a group of MBA interns from the Hult International Business School, Dubai campus. These include Srihari Chodagiri, Akhilesh Gupta, Rohit Kajaria, Sudhir Nijme, Kay Nikookary, and Tarun Talwar. At the start of the short internship period, all interns attended a three-day workshop on the methodology. Then we assigned each of the interns a project—a real project from SUKAD's internal portfolio—with the condition that they must follow the model. All of the interns did quite well and were able to demonstrate that there is value in the structured approach of CAMMP™, regardless of the type of project. The work of the interns and our internal work encouraged us to continue to enhance the approach, leading to Version 2, and now Version 3.

Version 2

The following professionals have been valuable contributors to Version 2 and the book that we published in 2014, *Redefining the Basics of Project Management*. They helped in reviewing the initial draft and include Ibrahim Awad, Sofien Dhouib, Theofanis Giotis, Youssef Saad, Luc Bauwmans, and Nada Chaban. A final mention to Professor Christophe Bredillet, who has been a supporter for many years and who also wrote the Foreword for the 2014 book.

Version 3 (current version)

Other professionals have provided input to the third version, the subject of this book. They contributed directly or via online discussions groups, and these include Bill Duncan, Mark Moore, Stephen W. Maas, Joel Carboni, and Neville Goedhals. Bill has convinced me to change the acronym of the methodology from CAM²P to CAMMP™ and the name of the first phase from Concept Phase to Discovery Phase. Neville was gracious enough to review a large section of the book. Dr. Joel Carboni has been a good friend and supporter; he reviewed the content related to sustainability and also provided the book Foreword.

Graphics Credit

A couple of years ago, I noticed the work of a university student and her sketchbook illustrations. We got in touch through social media, and I decided to use her work in one of my projects. Although this book is not the ideal place to show her skills, I believe it is still good enough to use sketches to represent the various images and graphical concepts. Sketches give this work a personal feel and also align with some of the concepts of this book, which are (1) a book by a practitioner for practitioners, and (2) as handwriting is unique to a person, so CAMMP™ must be specific to an organization.

Therefore, most of the graphical images—the hand sketches—are the work of a young and aspiring professional, Ms. Judy Abu Rustom. Judy has worked with the author to develop these images and to create an atmosphere of simplicity and clarity of the various concepts. Judy is a fresh graduate with a Masters Degree in Architecture, and she works out of Lebanon.

Acknowledgment for Publishing This Book

Finalizing this book for publication could not have been possible without the following:

John Wyzalek from Taylor & Francis, who accepted this book for publication under the Advanced Project Management Series, for his patience over the last year or so, including the need to rewrite a few chapters.

Theron Shreve of DerryField Publishing Services, who guided the transformation of the manuscript into the book that is in your hands.

Susan Culligan of DerryField Publishing Services, who has been patiently editing and typesetting my work, which I know is not easy, my being a non-native speaker of English. She has been wonderful guiding me through the editing process, including some substantial rewrites and re-edits.

About the Author

Mounir A. Ajam is an entrepreneur, author, speaker, coach, advisor, consultant, volunteer leader, and project management thought leader.

He is the author of *The Inheritance, a Story of Friendship, Community, and Project Management*; *Project Management Foundation*; *Redefining the Basics of Project Management*; *Applied Project Management*; *CAMMP™ for Mega Projects* (to be published), and a series of e-books, all on project management.

He is a senior executive with more than three decades of outstanding global and practical experience in capital project industries such as engineering, construction, petroleum, utilities, project management, and management consultancy. He has worked on projects worth billions of US dollars in North America, Europe, Southeast Asia, and West Asia. His experience includes working small and multiple projects and large and complex projects, including mega projects in the United States and in Southeast Asia.

Mr. Ajam is a co-founder and Chief Executive Officer of SUKAD Group, a leading project management provider with offices in Lebanon and the United Arab Emirates, founded in 2004.

Mr. Ajam and SUKAD play quite an active role in the project management community through various professional activities that are open to community members at no cost. He is heavily involved with the project management community at the regional and global levels. Globally, he has served in different roles and capacities, including serving on the Global Advisory Group to the Project Management Institute (PMI®) Registered Education Provider program (2005 to 2007) and as a judge for many PMI educational awards. He served on the 2008 PMI® EMEA (Europe–Middle East–Africa) Congress Project Action Team. He is also a graduate of the PMI Leadership Institute Master Class (Class of 2007).

In West Asia, Mr. Ajam served on the board of directors for the PMI chapter in the Arabian Gulf. He led the effort to establish a PMI chapter in the United Arab Emirates. He also led the effort and established the Global Project and Process Management Association (GPPMA). He served as GPPMA board chair for three years.

Mr. Ajam is an advocate of project management and recognizes its strategic value. He contributes to project management growth by publishing professional papers and articles on numerous platforms. These platforms included PMI Congresses, *Construction Week Magazine*, Dubai Quality Group, DKV Experts Channel, PMForum.com, Wamda.com, and other publications. He is the principal author of the SUKAD blog site (http://blog.sukad.com), in addition to a personal blog. In January 2017, the SUKAD blog has close to 300 articles and approximately 300,000 views from around the world.

For more information about Mr. Ajam, please visit his website at www.mounirajam.com.

About SUKAD

SUKAD Vision: **Project Management for All Aspects of Life!**
SUKAD Mission: **Be an Agent of Change and a Catalyst for Development!**

SUKAD was established in Dubai, United Arab Emirates, in 2004. In 2012, SUKAD opened another office in Lebanon. From these two offices, SUKAD has been providing services, mostly in West Asia and Africa, but also in Central and Southeast Asia. SUKAD is highly recognized as a leader in project management services, with a significant percent of revenues acquired through repeat business and referrals from leading organizations.

SUKAD has an extensive project management research and development program. Under the label and trademark The SUKAD Way™, this R&D effort has resulted in the development of proprietary products, including The Customizable and Adaptable Methodology for Managing Projects™ (CAMMP™) and The Seven Elements of Project Management Maturity™ (The 7Es™).

In addition to the PM Methodology and PM Maturity Model, SUKAD has developed numerous advanced courses and master certificates in project management. SUKAD has published a series of books, booklets, and sample projects in Arabic and English.

SUKAD is a corporate citizen and business with a heart. Over the years, SUKAD has provided numerous complimentary learning events to thousands of professionals either on our own (under our 2SPI™ program) or through partnerships with various organizations and universities.

In recognition of our business and community successes, in 2011/2012 SUKAD was recognized and ranked in the Dubai SME 100 ranking and the AllWorld Network Arabia 500 ranking.

For more information about SUKAD, please visit their website at www.sukad.com.

Section I, Part A
The Global Project Management Scene

Chapter 1

Definitions and Clarifications

1.1 Introduction

In project management online discussions and other platforms of communication, physical or virtual, one may find many misconceptions and disagreements, some of which are driven by lack of understanding of key terms, while others are due to inexperience or just different use in current practice. Therefore, in this first chapter, it is essential to clarify a few terms and offer the definitions of the terms that will be used extensively in the context of this book.

1.2 Standard

In practice, the word *standard* could mean many things. The dictionary definition is: "something used as a measure, norm, or model in comparative evaluations." Further clarifications:

- A standard can be anything that an organization establishes for its internal work, as a standard, for its staff to follow. In this scenario, the organization usually mandates the use of the standard. Some organizations might not mandate the strict compliance with a certain standard, however, if something is optional, then it is a guide or a guideline.
- A common use of the term standard usually refers to something (documents) a professional organization, or authority, publishes for others to use. For example, there are engineering standards, programming standards, and standards for numerous areas of practice.
- It is also common that some countries have standards-issuing authorities, which publish their standards or sanction the standards that others publish.
- It is also possible that some standards are adopted by the government—in this context, a standard would become a regulation (or code) mandated by law.

1.3 De Facto Standard

The *de facto standard* is a term that refers to a certain reference (standard, standard document, guide) that either practitioners of that domain commonly accept or is highly popular. This phrase is a marketing term, and just because such a term is used does not make the reference an officially accepted standard.

1.4 Framework

Dictionary definition: "basic structure underlying a system, concept, or text."

It is common to use the words *framework* and *guide* interchangeably, but that is not 100% accurate. A framework is a general guideline or an approach that an organization can adopt. The framework could include many components. For example, the *PMBOK® Guide* offers guidelines on how to develop a scope statement, a work breakdown structure, an estimate, and a communication plan. PMI defines the *PMBOK® Guide* as ". . . the scope of this guide is limited to the practice of project management" (Project Management Institute 2017, p. 3).

Other examples:

- ISO 21500 offers a framework for managing a project. However, because a standards organization issues this document, many also refer to ISO 21500 as a standard, although the official title is Guidance on Project Management.
- IPMA offers a competence framework. "The IPMA Competence Baseline is the common framework document that all IPMA Member Associations and Certification Bodies abide by to ensure that consistent and harmonized standards are applied" (IPMA 2015a).

To summarize, a project management framework offers suggestions for organizations on how to manage projects. These guides can be used to support a standard, and they can also be the foundation for building an *organizational project management system*.

Framework(s) could be the platform(s), foundations, or starting points that organizations use to build their organizational project management system!

1.5 Method and Methodology

It is common that some practitioners confuse the meaning of these two terms—method and methodology—and use them interchangeably. In general, a method is "a particular procedure for accomplishing or approaching something," whereas a methodology is "a system of methods used in a particular area of study or activity." In other words, a method is a specific way that is set or fixed, whereas a methodology is a wider term, "a system of methods."

Furthermore, a method is different from a framework because a method means there is a particular way of doing something—like systematic process, a step-by-step approach. Note that there may be more than one method of achieving a required outcome.

A typical project management method may follow a specific project life cycle or a similar approach. For example, Waterfall is a method with a project life cycle consisting of a few phases in sequence. SDLC refers to a Software Development Life Cycle with its phases. Some practitioners refer to Agile as a method, but it is an approach, more a framework, whereas Scrum might offer a particular method for software development projects.

To differentiate between method and methodology, please reflect on the following:

- PRINCE2® is a method: "PRINCE2 (Projects IN Controlled Environments) is a structured project management method" (AXELOS n.d.). Notice the use of the terms *structured* and *method* in the definition.
- PRiSM™ (from GPM) is a method: Projects integrating Sustainable Methods™ have specific deliverables and steps that one should apply to a project life cycle.
- SUKAD (Success Uniqueness Knowledge Attitude Development) developed a methodological approach that is founded on a project life cycle, which is the Customizable and Adaptable Methodology for Managing Projects™ (CAMMP™). Notice the use of the term *methodology* and not *method*, which is intentional. The reason for the utilization of the term *methodology* is that CAMMP™ offers an approach that is not rigid or fixed and has to be customized and adapted to the organizational and project context. The customized and adapted version, for a particular project's type, becomes a method.

> **The CAMMP™ Model**
>
> CAMMP™ is not a fixed method, it is not one-size-fits-all. It is an approach that can be used to generate various customized and adapted methods, in which each method is used for a given project based on the project environment, domain, or classification.

For example, managing a "small" real estate development project can follow an appropriate, *tailored approach*, whereas managing a large and complex real estate development project may follow *another method*. These two *methods* could have similar features, but with tailoring to a different level of size and complexity. In other words, CAMMP™ is "a system of methods," and the organizational project management system for an entity can and should include a set of methods, one for each project type.[1]

It is important to mention that it is *not* common to hear or read about specific project management methods in the professional community. This lack of mention is because methods are often custom-built (tailored)for an organization; they are internal resources, making them proprietary information.

1.6 Is a Method (Methodology) Also a Standard?

It can be.

If an organization officially adopts a methodological approach (systems of methods) or a particular method, that becomes its standard. For example, at SUKAD, CAMMP™ is the chosen

---

[1] In this context, *type* represents domain, sector or industry, and project classification.

standard. Some organizations have adopted PRINCE2®, PRiSM™, or Scrum as their methods. Others have built their proprietary methods. However, *method* and *standard* mean different things, as has been explained.

1.7 System

The term *system* may have more than one common connotation. For some, the term is popular for technology, such as a control system, a software system, or a project management information system.

In the context of this book, the term *system* refers to a project management system, whether on paper or in software, such as the organizational project management system (OPMS). This term is in line with the concept of system thinking, which is to think of the whole system, a 360° view. Therefore, the term OPMS refers to the policies, governance, methods, processes, and guidelines used in an organization to manage projects. One can also expand the definition to cover the management of programs and portfolios.

1.8 PM versus OPM

The most common use of the term *PM* (project management) is about managing a single project, whereas *OPM* refers to organizational project management, the organizational approach, framework, or system for managing the organizational projects. As mentioned earlier, a wider view of this would be the system for managing the organizational projects, programs, and portfolios.

Consequently, in the context of this work, PM refers to managing *a* project, and OPM refers to managing *the organization's* projects.

1.9 The Guide

In the context of this book, the term *Guide* typically refers to ISO 21500, whereas *PMBOK® Guide* is a reference to PMI's *A Guide to the Project Management Body of Knowledge®*.

1.10 Closing Comments

In closing this chapter, it is important to mention that there are many other terms requiring definition, and we will do so when appropriate in the book. Some of these topics are:

- The difference between project scope, stage scope, and product scope
- PBS and WBS, product and work breakdown structure, respectively
- The difference between product, program, and project
- Project life cycle, program life cycle, and product life cycle
- Gates, stage gates, and process gates
- Phases and stages
- Process groups, functions, knowledge areas, and subject areas

Chapter 2
Leading Global Associations

2.1 Introduction

Today, in the world of project management, there are numerous professional societies and associations advocating project management. Many of these are global organizations with national or local chapters or affiliated organizations. It is not possible to offer a detailed study of these associations, but a brief mention would be appropriate and necessary. To our knowledge, the associations mentioned here are non-profit organizations.

In the context of this work, there are two categories of associations. The first is made up of associations that advocate project management as a general domain, whereas the others advocate particular functions within project management, such as cost, quality, or risk.

2.2 Global Associations: General Project Management

The most recognized global associations focusing on project management as a general category are:

- **PMI, the Project Management Institute.** PMI is a centralized organization with global headquarters based in the United States and chapters in various countries and cities around the world. Individuals join PMI through the central body, and it is optional for them if they want to join one or more local chapter. PMI publishes numerous guides and standards. It also offers various certifications at different levels or specialties. The PMI framework for managing a project is *A Guide to the Project Management Body of Knowledge® (PMBOK® Guide)*, the emphasis of which is on project management processes and functions (knowledge areas).
- **IPMA®, the International Project Management Association.** IPMA is another global association, but it operates differently than PMI, since its membership consists of national

associations rather than individuals. IPMA publishes various standards and guidelines and offers certifications at different levels. The focus of IPMA is on competence rather than process, and it offers guides such as *Individual Competence Baseline* (ICB®) and *Organizational Competence Baseline* (OCB), among other resources.
- **GPM®, Green Project Management Global.** The focus of GPM is wider than its name indicates. It is not limited to "green" or environmental matters but expands to encompass the broader aspects of sustainability, which includes People (society), Planet (environment), and Prosperity (economy). GPM advocates the need to integrate sustainability considerations into project management and product development. The GPM standard for sustainability is the P5™ Standard. GPM also developed PRiSM™ (Projects integrating Sustainable Methods). Like PMI and IPMA, GPM also offers project management certifications at various levels, but with an emphasis on sustainability and project management.
- **GAPPS®, the Global Alliance for Projects Performance Standards.** GAPPS is different from PMI, IPMA, or GPM, as it is not open for general membership, nor does it have chapters or certifications. Its membership consists of an alliance of organizations and academic, industry, and professional associations. The focus of GAPPS is role-based standards, among other things. As of the time of writing, GAPPS has produced standards for project manager, program manager, and, most recently, project sponsor. Because of GAPPS's structure and focus, it is not well known in the general project management community.

2.3 Global Associations: Specialties

The most recognized associations focusing on particular topics of project management are:

- **AACE®, the Association for the Advancement of Cost Engineers International.** Although the name emphasizes cost engineering, this organization covers numerous areas, mostly related, but not limited, to cost and schedule management. AACE offers various certifications related to these topics as well.
- **SAVE®, the Society for the Advancement of Value Engineering International.** Similar to AACE, SAVE focuses on particular areas of project management, such as value engineering and value management.
- Other organizations (and initiatives) in the world advocate project management practices. Some of them could be private companies; others are industry or domain focused, such as for the software development or construction industries. These include the Scrum Alliance, the Agile Manifesto, the Construction Industry Institute (CII), the Guild for Project Controls, among numerous others.

2.4 Other Relevant Topics

In addition to professional associations, there are other initiatives or organizations worthy of mention:

- **ISO, International Organization for Standardization.** ISO is not a professional association but an organization that issue guidelines and standards on numerous topics from energy management, to quality management, to its 2012 guide for project management.
- **PRINCE2®** is a globally recognized method launched by the government of the United Kingdom. It has a set of project management certifications based on this method, and a private organization manages the certifications on behalf of the UK Government.
- Numerous other organizations offer significantly valuable resources for the domain of project management, from Japan to the USA and from the UK to India and Australia. Some of these are independent; others are part of IPMA. It is not possible to mention them all here as they are so numerous, some of them have overlapping content, and the author has not had the opportunity to study them all.

2.5 Closing Comments

Reflecting on the above, one may notice that these associations could supplement each other, because their areas of focus are quite diverse. However, it is unfortunate that, at times, they compete rather than collaborate.

Project management is a vital contributor to organizational performance!

Later in the book, the author makes the case that these associations do indeed offer complimentary products, and the offered solution covered in Section II of this book integrates some of what these associations offer. The Customizable and Adaptable Methodology for Managing Projects™ include elements from these various associations. CAMMP™ endeavors to integrate this knowledge and transform concepts into a practical, comprehensive, systematic, and adaptive methodological approach.

The aim is to encourage organizations (governmental, non-governmental, for-profit, and not-for-profit) wanting to deliver projects successfully and reach the highest level of organizational project management maturity, and these entities should consider integrating the learning from these associations to maximize their chances of success and realize the benefits of their initiatives.

Project management is a vital contributor to organizational performance. CAMMP™ and this book are starting points; where to go from here is a challenge to organizations, which they can convert into opportunities for continual success.

Chapter 3

PMBOK® Guide Overview

3.1 Introduction

A Guide to the Project Management Body of Knowledge® (*PMBOK*® *Guide*) can be called the flagship publication of the Project Management Institute (PMI); it is a project management reference with an extensive global distribution. Its content summarizes many concepts obtained from decades of professional practice by the volunteers who wrote it and regularly update it. It also captures the knowledge from various other resources. Therefore, it would be proper to briefly review the *PMBOK*® *Guide* and discuss some of the myths about it, which were contributing factors to developing The Customizable and Adaptable Methodology for Managing Projects™ (CAMMP™).

Consequently, some of the content of this chapter (and this book) might be a shock to the conventional wisdom of a few project management practitioners. Will some readers experience a paradigm shift? Let's see!

3.2 Brief History

The first official, consolidated copy of the *PMBOK*® *Guide* was published in 1996. It was about 180 pages; it had 37 processes, nine knowledge areas, and five process groups. This first edition did not have an ANSI standard. Since 1996, there has been a new and updated edition of the *PMBOK*® *Guide* every four years, with the latest officially published edition being the sixth edition, released in September of 2017.

The major changes to the *PMBOK*® *Guide* since the original version are:

1. Part of the *PMBOK*® *Guide* was recognized by ANSI, the American National Standards Institute, as an official ANSI standard. The third edition (2004) and fourth edition (2008) included this standard in Chapter 3. it became an Annex in the fifth edition (2013), and in the sixth edition it is Part 2.

2. With the fifth edition, the addition of one knowledge area—Project Stakeholder Management—aligns the guide with ISO 21500. With the sixth edition, there are further alignments with ISO, in particular changing *time* management to *schedule* management, and *human resource* management to *resource* management.
3. In every update, the *PMBOK® Guide* included changes to the number of processes, up or down, moving some processes from one area to another or one process group to another, occasionally changing the name of one or more processes.

Table 3.1 summarizes the general changes from the first to the sixth editions.

Table 3.1 Summary of Some of the *PMBOK® Guide* Changes

Year	Edition	PG[a]	KA[b]	Processes	Pages[c]
1996	1996	5	9	37	~ 180
2000	2000	5	9	39	~ 210
2004	Third	5	9	44	~ 400
2008	Fourth	5	9	42	~ 460
2013	Fifth	5	10	47	~ 620
2017	Sixth	5	10	49	~ 790

[a] Process groups
[b] Knowledge areas
[c] Cover to cover

At the time of finalizing this book, PMI has just published the sixth edition. Suffice it to say that the changes from the fifth to the sixth edition do not have an impact on this book. There might be minor adjustments, but the core concepts are unchanged, which means the core message of this book will still be applicable and not affected.

3.3 Relevant Highlights from the *PMBOK® Guide*

3.3.1 Process Groups

Following are the key notes about the process groups and processes.

> There are five process groups, consisting of 49 processes.

- The *PMBOK® Guide* includes five process groups.
- Each process group consists of two or more processes, with the exception of the closing process group, which has only one process.
- In total, there are 49 processes, per the sixth edition.
- Over the years, with the different versions, the number of process groups has not changed, whereas the number of processes typically changes from one edition to another.
- The five process groups are for *a project or a phase*.

3.3.2 Knowledge Areas

> There are ten knowledge areas, consisting of 49 processes.

There are 49 processes distributed across the various knowledge areas, based on their focus.

- In the sixth edition, there are 10 knowledge areas.
- The number of knowledge areas has had not changed over time until the fifth edition, when the number changed from nine to ten, adding the chapter on stakeholders.
- Each knowledge area consists of a few processes.
- Therefore, *every process belongs to a knowledge area and a process group*.

For a list of all of the *PMBOK® Guide* processes and their mapping to a process group and knowledge area, please refer to the sixth edition, Table 1-4, p. 25.

3.3.3 Generic Standard Document

The *PMBOK® Guide* is generic in nature and is useful for "most projects, most of the time" (The Project Management Institute 2017). It is not tailored or customized to a particular sector, industry, domain, or project type. Because of the generic nature of the guide, PMI has published a few industry-specific extensions (supplements) to the *PMBOK® Guide*.

In addition to the industry/sector extensions, there are numerous standard documents for special topics such as work breakdown structure, estimating, scheduling, and risk management. PMI labels them *practice standards*. Furthermore, there are standards for program and portfolio management and organizational project management.

3.4 Myths about the *PMBOK®* Guide

There are many myths about the *PMBOK® Guide*, but we will focus on only the two that are relevant to the context of this book.

3.4.1 The PMBOK® Guide Is About Best Practices

Marketing drives a common misconception about the *PMBOK® Guide*, since many educational providers promote it as *project management best practices*.

A search through the *PMBOK® Guide* for the phrase *best practices* brings back a few hits, but most are in relation to *industry best practices*—*practices from outside the guide*. Nowhere was the term *best practices* used to refer to the *PMBOK® Guide* itself. On the contrary, the definition and purpose of the *PMBOK® Guide*, as stated in its Chapter 1, mentions *good practice*. In other words, because of the generic nature of the *PMBOK® Guide*, it can only offer *good* practice or *common* practice, whereas best practices would be specific to an industry or domain.

Why Is This Relevant?
If the *PMBOK® Guide* is THE BEST PRACTICES, then why look elsewhere? Can anything top THE BEST?

These questions are not rhetorical statements; they have a direct impact on project performance if practitioners do not seek knowledge from other sources.

In Section II, Part F of this book (The Third Dimension), includes a chapter on best practices, which provides further clarification on this myth.

The *PMBOK® Guide* is about good practice;
it does not offer project management best practices!

3.4.2 The PMBOK® Guide Is a Methodology

Marketing partially drives this myth as well, but its cause is mainly due to a genuine misunderstanding of the intention or reach of the *PMBOK® Guide*, including by those who deliver classes about it.

It is quite common to see advertisements for courses, request for proposals, or online discussions referring to the PMI, PMP®, or the *PMBOK® Guide* Methodology. Well, one does not exist. The PMP is a professional certificate, the *PMBOK® Guide* is a framework, and PMI does not offer or promote a methodology.

It is vital to repeat something covered earlier: the *PMBOK® Guide* is generic—its popularity is in being generic and "for most projects most of the time." On the other hand, a methodology or method has to be tailored, custom-fit to organizational needs and the project environment. Therefore, the *PMBOK® Guide* is not and cannot be a methodology.

Chapter 1 of the PMBOK® Guide is clear about this fact.

Consequently, because the *PMBOK® Guide* is not a method, the recommended practice is for organizations to develop their methods, which can align to the *PMBOK® Guide*, which is what SUKAD did in 2007 with the development of CAMMP™.

Once Again, Why Is This Relevant?

This is relevant because it helps project management practitioners understand that

> **Estimating Example**
>
> Here is an example to elaborate on the concept of the *PMBOK® Guide* and methodology. In the chapter on cost, the *PMBOK® Guide* covers the cost processes, such as estimating and budgeting.
>
> It elaborates on estimating techniques, but it does not explain well where or how to use the different techniques. Furthermore, it does not explain the number of estimates required on a given project. On the other hand, a method will define the number of estimates required along a predefined project life cycle.
>
> Furthermore, the *PMBOK® Guide* presents budgeting, but it does not elaborate on when a budget is approved or who approves it and at what stage gate. A method will clearly establish all of these points.

the *PMBOK® Guide*, on its own, is not enough to manage projects—*there is a need for a method to supplement it*. Without a method or a well-defined project life cycle, the management of projects is deficient. A critical point related to this myth is the confusion between process groups and project phases, which is the subject of another chapter.

3.5 Closing Comments

Except for the addition of the chapter on stakeholders in the fifth edition, there have been no significant *structural* changes to the *PMBOK® Guide*. Yes, the number of processes increased and decreased, other processes changed names or location, and content was added here and there. However, a significant change to highlight here was that part of the guide became an official ANSI standard with the third edition. It is important to note that the sixth edition has introduced more discussions on Agile and tailoring and replaced the *human resource* management with *resource* management, but still no major structural changes. In the author's view, the guide should be completely re-structured and re-written, but that is outside the scope of this book.

It is also important to stress the discussed myths for a better understanding of the guide—what it is, and what it is not. It is crucial to see the big picture and know the concepts and principles and how to apply them correctly. It is regrettable that many practitioners stop at applying part of the *PMBOK® Guide* only, not recognizing how deficient their practices are. The unfortunate reality is that these practitioners claim to implement the *methodology* and *best practices* per the PMI "Global Standard."

Since the author is not a position to modify the *PMBOK® Guide,* this book can serve as a practical guide that provides the reader with a comprehensive approach for applying innovative and efficient project management. Consequently, it would be critical for individuals and organizations to think out of the box, see the big picture, as they build methods and systems for applying project management.

It seems that shifting people's paradigm would be vital to delivering exceptional organizational value.

Chapter 4

ISO 21500 Overview

4.1 Introduction

ISO 21500 is the reference for guidance on project management. It is the first guide by ISO specific to project management. In this chapter, we will highlight the parts of ISO 21500 that are *relevant to the offered solution and the proposed methodological approach*. ISO 21500 is a vital reference for this book and The Customizable and Adaptable Methodology for Managing Projects™ (CAMMP™).

4.2 Development and Influence

Various standard organizations from around the world participated in the development of ISO 21500. The influence of the Project Management Institute (PMI) and the International Project Management Association (IPMA) is present in the ISO guidance document as an effort to ensure some degree of alignment. For example, both the *PMBOK® Guide* and ISO 21500 share the concept of process groups and knowledge areas, what ISO 21500 calls *subject groups*. ISO 21500 also includes discussions of competencies of project personnel, which is an IPMA influence.

Once ISO released its guide, it was time for the professional associations to align to ISO 21500. For example, the fifth edition of the *PMBOK® Guide*, which PMI released after ISO 21500, included a new knowledge area on stakeholder management, which is an ISO subject area. The sixth edition of the *PMBOK® Guide* includes other changes that align to ISO 21500, such as replacing the human resource chapter with one on resource management, which is also an ISO 21500 subject area.

4.3 The Process Groups

The ISO *Guidance on Project Management* presents the concept of process groups. The ISO 21500 process groups are initiating, planning, implementing, controlling, and closing.

18 Project Management beyond Waterfall and Agile

The process groups' origin is founded in the principles of total quality management, in particular, the Deming Cycle of PDCA (Plan, Do, Check, Act), as shown in Figure 4.1. Linking the PDCA to process groups, one can notice how ISO 21500 transformed the PDCA Cycle into the process groups.

- The Planning Process Group represents *Plan*.
- The Implementing Process Group is used instead of *Do*.
- The Controlling Process Group replaces *Check* and *Act*. (It is not clear why ISO combined these two quality steps into one process group.)

Figure 4.1   The PDCA Cycle (Plan-Do-Check-Act).

It is worth noting that the PDCA Cycle originated in the operating, manufacturing environment, in which the operational activities are ongoing. However, projects are temporary, meaning they must have a starting point and an ending point. Therefore, it was important to add *Initiate* and *Close* to complete the picture and transform the PDCA Cycle into the project management process groups (see Figure 4.2).

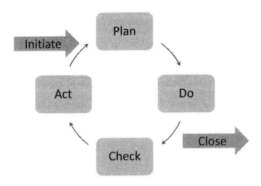

Figure 4.2   The PDCA Cycle with Initiate and Close.

4.4 Subject Areas

ISO 21500 refers to the project management functions, what the *PMBOK® Guide* calls *knowledge areas,* as *subjects.* There are ten subjects: integration, stakeholders, scope, cost, time, quality, risk, communication, resources, and procurement.

4.5 Processes

ISO 21500 presents us with 39 project management processes. There is no need to list them at this time; please refer to the Guide for a full list.

What is important to emphasize here is that ISO 21500 states that there are three categories of project processes required for projects:

- Project management processes
- Product processes
- Support processes

Although ISO 21500 mentions all three categories, its focus remains on the project management processes.

Three categories of project processes: project management processes, product processes, and support processes

---

4.6 ISO versus the *PMBOK® Guide*

At this point, it is important to clarify that technically, *it would not be appropriate to compare ISO 21500 to the* PMBOK® Guide—the full guide—because they are not equivalent.

The *PMBOK® Guide,* as a book, includes the ANSI-approved project management standard advocated by PMI. This standard is Part 2 in the guide, and it is less than 100 pages. The rest of the book is the guide, which includes elaborations on the standard. Therefore, for a proper comparison, one should compare ISO 21500 to Part 2 of the *PMBOK® Guide* sixth edition and not the full guide.

4.7 Closing Comments

ISO 21500 is similar to the ANSI Standard; it does not cover best practices, nor is it a method. It provides the processes that can be used for managing projects.

Throughout this book, any reference to the process groups is in reference to ISO 21500. In Section II, Part D, the author proposes an alternative approach to the process groups.

Simplistically, we could treat ISO 21500 and the ANSI Standard as equivalent, at least in the material that is relevant to this book on CAMMP™. The decision to reference both is because both are valuable resources, and each has its practitioners.

Chapter 5
Other Relevant PM Standards

5.1 Introduction

As outlined in Chapter 1, there are numerous professional associations, each of which publishes a set of standards and guides. This chapter is concerned with two of those organizations and some of what they publish. The third dimension of The Customizable and Adaptable Methodology for Managing Projects™ (CAMMP™) includes topics related to the associations covered in this chapter; these topics are covered in more detail in Section II, Part F.

5.2 IPMA®

Over the years, a leading IPMA® (International Project Management Association) reference has been the IPMA Competence Baseline (ICB®), version 3.0 of which was in place until 2015. In 2015, IPMA restructured the ICB and published ICB 4.0, but this time the "I" in ICB was changed to represent *Individual,* not *IPMA,* and ICB 4.0 become the Individual Competence Baseline. With this shift, IPMA also published the Organizational Competence Baseline (OCB®) and other references, in line with their Project Excellence Model.

For the purposes of this work, our interest is related to ICB, since it is relevant to the third dimension of the CAMMP™ Model. "The IPMA Individual Competence Baseline (ICB) is the global standard for individual competence in project, programme[1] and portfolio management"; ". . . ICB is not a 'how-to' guide for managing projects . . . therefore, it does not describe the processes or steps involved in project, program or portfolio management."[2]

ICB offers the professional community three competence areas:

---

[1] IPMA uses UK–European English, not North American English, hence the difference in the spelling of some terms.
[2] IPMA, International Project Management Association, 2015.

- People competences, such as personal and interpersonal competencies
- Practice competences, specific to projects, program, or portfolios
- Perspective competences, which would be more about the environment "as well as the rationale that leads people, organisations, and societies to start and support projects, programmes, and portfolios."[3]

In total, there are 29 elements of competence.

While ISO and PMI offer the project management community processes, process groups, and subject areas, IPMA offers the community the competence elements to enable the successful completion of the projects' work.

5.3 GPM® Global

GPM® (Green Project Management) is a relatively young organization with a specialized focus on sustainable project delivery. GPM has published a P5™ Standard that addresses the five Ps of People, Prosperity, Planet, Product, and Process. In other words, GPM advocates that as organizations work on a project, they consider the project impact on people (society, workers) and the planet (environment), along with their prosperity (economic) considerations.

To translate the P5 Standard into a practical approach, GPM published a guide and a method under the name of PRiSM™, Projects integrating Sustainable Methods. PRiSM is built on the ISO 21500 platform as its project management foundation, and adds the sustainable concepts and deliverables along the project life cycle. The gap in PRiSM is that it depends on ISO process groups as the project life cycle, a gap GPM is working on closing as they update their guide. As it stands today, the author views PRiSM and its concept as a layer in the third dimension of the proposed methodological approach (refer to Part F).

5.4 Closing Part A

In closing Part A of this book, it is important to link it to upcoming parts. This part focused on the global project management scene and the current reality.

The core of this book, six out of eight parts, is about the CAMMP™ Model, which is a three-dimensional model that integrates knowledge from these various associations.

- The first dimension reflects a project life cycle model that SUKAD developed based on the author's decades of experience in the capital projects industry.
- The second dimension incorporates the project management processes as modified by CAMMP™.
- The third dimension considers numerous resources, but two essential components are competence from IPMA and sustainability from GPM.

Part B highlights the challenges of the current reality and practices.

---

[3] Ibid.

# Section I, Part B
The Challenges in Current Practices

Chapter 6

Gaps and Opportunities

6.1 Summary of Previous Chapters

To summarize the relevant information from the earlier chapters, the current practice is:

- PMI and ISO are clear that they are not offering the community a method or methodology. They are providing a set of processes, project management process groups, and subject/knowledge areas. ISO 21500 mentions the need for product and support processes but does not address them.
- IPMA is also clear that it does not offer "how-to's"; rather, it advocates the competence elements for managing projects. Here again, there is no method.
- GPM offers a method, but although its dependence on the process groups as a project life cycle is a weakness, its sustainability elements are of great value.
- It is important to state that PRINCE2® is a method, which is good; but for some reason, it is mostly known in the UK and other countries with organizations that have a UK influence. The author does not offer a dedicated chapter to PRINCE2, because CAMMP™ is an alternate solution that is more flexible and wider in scope.

6.2 Transition, Understanding the Challenges

The hypothesis of this book is that, despite the high value each professional association offers, there are still gaps in project management practice. Practitioners still struggle to apply what they learn in the real world, on real projects, and on different types and classes of projects. In the world of projects and project management, certain fixed concepts apply regardless of industry or domain. Many variables are highly unique to the context of a given project.

Yes, organizations can use the IPMA's ICB® and develop their methods using the competence elements.

Yes, organizations can use the process groups and subject groups from PMI/ISO to develop an internal methodological approach.

Some are doing so, but not enough!

In large organizations with abundant resources, their staff could explore the world of project management and choose what is best for their organizations from the available "menu" of options. Even in such organizations, one can find that they stick to one menu item, or one resource, for one reason or another.

While large organizations may limit their choices, small and medium organizations may not even have the luxury of selection. Consequently, they constrain their project management system—assuming they have one—and depend on the common sense of their accidental project managers. These organizations manage projects, or, more accurately, "execute" projects through accidental project managers, then wonder why the failure rate is so high. It is also possible that these organizations think that they are delivering the project successfully; this might be so, but are they using clear criteria for measuring project success?

These practice gaps exist because organizations tend to box themselves into limited options. The gaps present us with opportunities to provide workable solutions. The fundamental principle of the offered solution revolves around integrating the best of what exists and offering it in a practical approach that can work for small or mega projects, regardless of domain, type, or class of project. This is a modest attempt to save organizations much research and development work.

The ultimate objective is to help project management practitioners enhance their own and their organizations projects' performance.

6.3 Objectives

Therefore, the objectives of CAMMP™ and this book are to offer individuals and organizations a new way of thinking, a simplified yet effective approach for *applying the principles of project management on real projects in the real world.* The new thinking integrates the following concepts:

- Emphasize that the process groups are not the project life cycle, and they are not the project phases.
- Offer a new perspective on the process groups, specifically the planning process group.
- Emphasize that those standards and guides from professional associations must be part of a holistic organizational project management system, not a standalone resource.
- Offer ideas on how to apply principles of global standards in the real world.

> **Could It Be This Bad?**
>
> A study by PricewaterhouseCoopers, which reviewed 10,640 projects from 200 companies in 30 countries and across various industries, found that only 2.5% of the companies successfully completed 100% of their projects (Gallup 2016).

- Introduce a proposed three-dimensional methodological approach for managing projects.
- Provide ideas on how to tailor the three-dimensional model and apply for different types of projects.

6.4 A New Perspective or Shifting Paradigms?

What the author offers in this book is a different viewpoint. Some content may even challenge the conventional wisdom for some readers and may lead to shifting their paradigm. Some readers might consider part of the content as critique, or even criticism, of one guide or another. However, a careful reading of the offered solution may lead practitioners to think outside the box.

Must think out of the box, without ignoring the box!

What is offered here is a professional subject-matter expert's opinion based on real-world practice on small, large, and mega projects. These concepts have been applied in industrial projects as well as on day-to-day internal projects—projects such as writing and publishing a book; launching a new office; launching a business; developing a web portal; building a petrochemical facility, research, and development program; an online application project; and many others.

The bottom line: growth and continual improvement of any system cannot happen without developers and practitioners being open in addressing weaknesses, gaps, and challenges.

Chapter 7

Clearing a Critical Challenge

7.1 Introduction

This chapter is critical to the message of this book and is one of the core reasons for developing The Customizable and Adaptable Methodology for Managing Projects™ (CAMMP™) and writing this book. The challenges that this chapter address are the most common areas of confusion in the project management community among practitioners who subscribe to process groups concepts, which are addressed in the *PMBOK® Guide* and ISO 21500. Numerous practitioners, certificate holders, education providers, and "trainers/instructors" do not fully understand these guides and what is explained here.

All it takes to prove this point is to go online to any social media group related to project management and ask people to name the phases of a project—and read their answers.

In this chapter, the text often includes reference to the names of process groups. The names used are per ISO 21500.

7.2 The Critical Challenge

7.2.1 Process Groups and Project Life Cycle

One of the critical gaps in practice is the confusion between the *process groups* and *project life cycle*. The cause of the confusion is not an error in the guides, as both the *PMBOK® Guide* and ISO 21500 are clear that the process groups are *not* project phases. The guides also stress that the project life cycle is a variable, a function of domain and industry, whereas the process groups are the same across different types of projects.

Many practitioners and students of project management misunderstand these guides, and they apply what they *believe* they understand, which results in less than optimal solutions. It is

30 Project Management beyond Waterfall and Agile

unfortunate that some of those who fail are now blaming the Waterfall methodology, claiming that Agile methodology is the way to go. This blaming practice shows a significant lack of professional understanding, since neither ISO 21500 nor the *PMBOK® Guide* is about the Waterfall method, and Agile is an approach rather than a methodology.

Back to the point: when one asks these practitioners to name the phases of the project life cycle, they provide the name of the process groups as stages, per Figure 7.1.

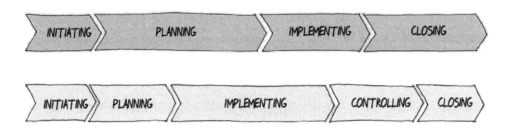

Figure 7.1 How some practitioners of project management understand the project life cycle.

The top part of Figure 7.1 presents one view of some practitioners who think that the project life cycle is fixed and consists of four phases, which are four of the five process groups. In the bottom part, there is the addition of *Controlling*, since some of these practitioners either think this is a phase or just confuse the process groups (all of them) as phases of the project life cycle. In other words, for this second group, *Controlling is after Implementing,* with no control actions until this point in the project life cycle.

These practitioners do not recognize two fundamental facts:

- The *process groups ARE NOT project phases.*
- These *process groups repeat* in every phase.

The unfortunate reality is that the situation is not limited to individuals, as even some organizations are labeling their projects' phases after the process groups. Therefore, it is not uncommon today to see organizations that have initiating, planning, implementing (or executing), and closing phases.

A planning phase might require a short span out of the overall project time span, whereas the planning processes will span most of the project life cycle!

The issue is not the names and terminology. It is not a problem with whatever names individuals and organizations use for their project phases, as long as they understand that a *planning phase* is different from the *planning processes,* and the *implementation phase* is not the same as *implementing processes*. For example, a planning phase (if one exists) might require a short span out of the overall project time span; while the planning processes will span most of the project life cycle as the work progresses from one phase to another.

Again, names are not critical as long as one differentiates phases from process groups. This distinction is a crucial matter, since misunderstanding the distinction will have direct negative consequences on project performance.

In other words, the confusion is not limited to terminology, but to the practice.

7.2.2 Opinion or Fact?

Another unfortunate situation occurs when one presents the above clarification—professionals think this is only an opinion and that it is not logical. So, is this an opinion or a fact?

In case there is still some doubt, here are references from the sixth edition of the *PMBOK® Guide*.

- The *PMBOK® Guide* is clear that the project life cycle consists of phases (not process groups). (See Part 2, Section 1.9, page 555.)
- The *PMBOK® Guide* also states that the process groups repeat in every phase. (As above, see Part 2, Section 1.9, page 555.)

Similarly, ISO 21500 states, "Each process group consists of processes that are applicable to any project phase or project," Section 4 (ISO).

Some readers are thinking at this time, "We know this," and many do. However, when it comes to applying this concept, some of them stumble and fall back into thinking the *process groups* are *phases*.

For example, in response to the question: "How many charters are there along the project life cycle?" it is likely that some practitioners answer "Only

> **Terminology**
>
> The names of some of the *PMBOK® Guide* and ISO 21500 processes contain the word *project* in the name. This nomenclature results in a contradiction between the text in the guides (where the processes are explained) and the names of the processes.
>
> For example, what does the Develop Project Charter process mean? How about Develop Project Management Plan and other similar processes with the word project in the name? Are these processes applicable to the phase or the project? If these processes are for the project or phase, as is often repeated in the text, why do the names of these processes include the word *project*?
>
> For example, when professionals read Develop Project Management Plan, they think *project*, not *phase*. When they read Direct and Manage Project Work, they think *project*, not *phase*.
>
> This nomenclature creates significant confusion, and practitioners think the process groups apply once on the project level and, therefore, they are the project phases. They believe there is one charter that is fixed and never changes. They also believe that there is one management plan with subsidiary plans from the knowledge areas; there is one estimate, one schedule, one procurement, one team, one WBS, and so on.
>
> This point is not about terminology preferences, but rather its critical role in creating confusion, which results in applying this concept to deficiencies, shortcuts, or improper practice.
>
> Consequently, projects suffer and performance is less than optimal, if not mediocre or a failure.

one." Well, if the process groups repeat—including the initiating processes, which in turn include the charter process—does that not mean that the *Develop Charter Process* repeats, and that there is a charter for every phase?

7.2.3 So, Why the Confusion?

If ISO 21500 and the *PMBOK® Guide* present that process groups are not project phases, then why do so many practitioners and project management professionals miss this point?

Why are even organizations confusing project phases with process groups?

The following are possible explanations:

- Many do not read the guides carefully. If they do, they focus on processes, input, tools and techniques, and output.
- The focus on the processes, process groups, and subject areas often lead a person away from the sections where the Guide presents the concept of a project life cycle.
- The terminology inconsistency in the guides is another clear contributor, which is the use of the word *project* in the name of processes (refer to the terminology inset).

There are many other reasons, but perhaps it is not politically correct to state them here. The unfortunate thing is that there are various publications and posts that help in spreading this confusion. There is not a single day or week that goes by without a blog or an online post that considers the process groups as project phases.

Next, it is important to define all of these terms and present how to integrate the process groups within the project life cycle.

7.3 The Project Life Cycle

7.3.1 Project Life Cycle Definition

What is the project life cycle?

Here are a few points to define it:

- The simple answer is that the project life cycle is a span of time from the start to the end of a given project. However, determining where the beginning and end are is not as simple as it sounds.
- Some project management practitioners use the term *project life span* for this concept.
- A typical project life cycle consists of a few phases or stages.

> It is important to recognize that a software development life cycle (SDLC) is a specialized life cycle specific to software development. It is not a project life cycle that applies in all situations.

7.3.2 Other Life Cycles

It is critical to distinguish between *project life cycle* and other life cycles such as *product* life cycle and *program* life cycle.

- A *product life cycle* is the life of a product (building, hospital, software, medicine), which starts with the concept of launching the product and goes through phases. The product life cycle phases usually include acquisition, operation, and maintenance. The product life cycle ends with taking the product off the market. During the product life, there can be numerous projects and programs, each with its own life cycle. A product life cycle length may vary from about three years for a software project, five to seven years for an automobile, or twenty-five years (or more) for a facility project.
- A *program life cycle* is the span of time—start to finish—of a program. A program typically includes numerous projects that can be in sequence or with some overlaps. Therefore, during the program life, there are many projects, each project with its own project life cycle. Programs are typically linked to strategic goals and may last for many years.
- There could be other concepts that include the term *life cycle*, but they are not directly relevant here. Suffice it to say that the term *life cycle* is used to represent a process, end to end. It could be a change, a purchase order, a contract, or other actions.
- Finally, a *project life cycle* must be peculiar to a given project and *not* a sub-project, a program, or a product. It is also important to recognize that a project could be part of a program or independent of one.

7.3.3 Project Life Cycle Is a Variable

In project management, some things are domain specific and vary from one domain to another; the project life cycle is one of those

> **Sub-Projects**
>
> The term *sub-project* usually refers to a part of the project that could be split out for one reason or another.
>
> For example: Block A, Block B of a project that consists of multiple blocks.
>
> The split could be just a way of distributing the workload or for other reasons. The sub-project could be within a phase or have multi-phases.
>
> Back to the multi-block examples: All blocks could progress together as one project until the team reaches construction, at which time they decide to split the work between multiple contractors, one contractor per block.
>
> Alternatively, each block could be treated as a sub-project, with various stages such as design, procure, build—each block independent of the others.
>
> Why would these be considered sub-projects (within a project) instead of projects in a program? One reason is approval and funding: if these sub-projects are justified and approved together, then they would be a project.

things, whereas there are things that cross domains, such as a given process or process group.

For example, the project life cycle is highly dependent on the industry and project domain. Therefore, the number of phases, their names, and other life-cycle elements vary from one domain to another. Consequently, there is no fixed project life cycle for all projects. Nonetheless, within the same domain, the project life cycle could be similar from one company to another—at least the main phases.

However, it is also important to consider the perspective of the organization when one discusses life cycles, and this is next.

7.3.4 Whose Perspective?

The previous definition is generic and can lead to differences of opinions. Therefore, one cannot answer the question, "What is the project life cycle," without having clarity about the scenario.

- For example, who wants to know?
- From whose perspective is the question being asked?
- What is the background of the person asking?
- What type of organization does this person represent?

This point is important, because the word *project* and the phrase *project life cycle* can mean different things to different people; it all depends on how one relates to the scenario. As a result, before one can answer the life-cycle question, one needs to determine if the questioner is a service provider or a project owner.

For a project owner organization, the project life cycle could be from idea or authorization to closure, whereas a service provider could be involved in only one phase of the project owner's project life cycle, such as concept phase, design, or construction.

More on project life cycle in Part E of this book.

7.4 Project Phases and Stages

First, a reminder: The project life cycle represents a time span from the start (idea, business case, authorization) to the end (completion and acceptance). This time span could be weeks for small, simple projects or years for large, complex projects.

For better control of the project output and outcome, it is common to divide the project life cycle into shorter spans, or *time segments*. While it is common to refer to these time segments as *phases*, others refer to them as *stages*. It is common to use the terms *phases* and *stages* interchangeably, but for CAMMP™, phases and stages have similar characteristics but represent different things.

7.5 Clearing the Confusion

It is time to clear the confusion with a visual demonstration, mapping the process groups to a project life-cycle model. Just a reminder of a few important concepts:

- The project life cycle, in general, is a variable; it changes from one domain to another and could vary even for the same type of projects.
- The project life cycle usually consists of three to four phases.
- The process groups are "fixed" and applicable, regardless of the type of project or domain.
- The process groups repeat for every phase.

7.5.1 Phase Perspective

The process groups occur in every phase, but they are not phases themselves. This statement is worth repeating because of its importance. In other words, apply the process groups *within* a phase.

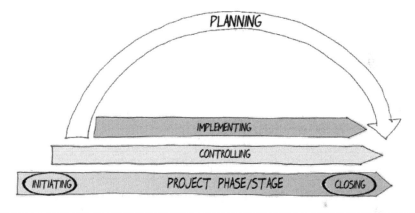

Figure 7.2 The process groups within a typical project phase/stage.

Figure 7.2 presents the process groups for *a typical project phase/stage;* the following is a brief explanation:

- Notice the *Initiating* in the oval at the start of the stage, which indicates initiating the *stage*.
- Then, there is an arch for *Planning,* indicating planning the *stage*. The reason it is an arch (like an umbrella) is to indicate the overall perspective, and that the plan should cover the stage work, end-to-end. Furthermore, in addition to the stage plan, there could be other planning activities throughout the stage. The plan should be completed shortly after initiating the stage.
- Then there is a horizontal arrow for *Implementing,* indicating implementing the *stage* work. Implementing work starts once the plan is approved and ends just before closure.
- Another horizontal arrow represents *Controlling* during the *stage*. Notice, this arrow is longer than the implementing arrow, it starts before the beginning of implementing and ends after it. The length of the arrow reflects that control touches all process groups and begins with initiating and ends with closing. This is important because some practitioners think that control is only implemented during execution and after plan completion.
- Finally, *Closing* on the right shows the closure of the *stage*.

7.5.2 Project Perspective

It is time to review the process groups in comparison to the project life cycle. *Imagine the project as one phase while reading the next few paragraphs.*

Initiating

Figure 7.3 presents an eight-stage project life cycle. The stages are shown linearly here, although some of them may overlap slightly or significantly. This illustration shows *Initiating* in the oval on the left side of the image, indicating the project's initiation.

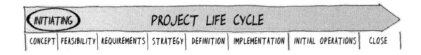

Figure 7.3 A project life cycle, with the initiating process group.

Planning

After initiating the *project,* it is time to plan the *whole project.*

Planning the entire project is represented by the *Planning* arch, as shown in Figure 7.4.

This plan is a general, *high-level plan* with a *limited amount of detail;* think of it as a project management strategy for taking the project to the end and delivering success. Let us emphasize the points and elaborate:

- First, this is the plan for the whole project, not a stage.
- It is high level because there are not enough details at this early time in the project life cycle, at least for the subsequent stages.
- Each of the upcoming stages will have its own *stage management plan*.
- Since the project management plan takes the perspective of the whole project, notice that the arch stretches to the very end of the project life cycle.

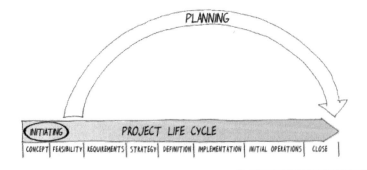

Figure 7.4 A project life cycle, with the planning process group.

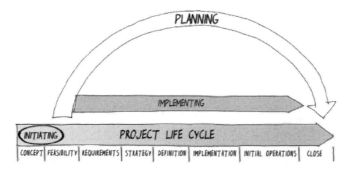

Figure 7.5 A project life cycle, with the implementing process group.

Implementing

In Figure 7.5, the implementing process group is added.
Here are a few points to highlight:

- Notice that *Implementing* the project in this graphic goes through most of the stages; this is the horizontal arrow spanning to the end.
- Therefore, *Implementing* the project starts after initial planning and stops just before final closure.

What this means is that *implementing the project is going through the stages:* requirements, strategy, definition, implementation, initial operations, and even part of the Close Stage.

Controlling

In Figure 7.6, the controlling process group is added.

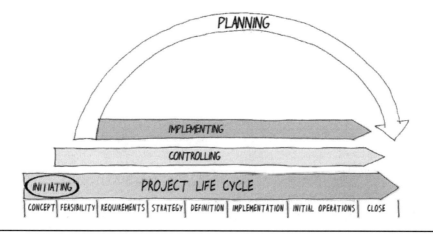

Figure 7.6 A project life cycle, adding the controlling process group.

Here are the highlights for monitoring and controlling:

- Notice that the *Controlling* arrow goes through all the other stages, spanning to the end, and is longer than the *Implementing* arrow.
- What this means is that controlling processes start as soon as the project is initiated and do not end until just before the project close-out report.
- In other words, controlling processes are not only in relation to the project management plan; the project is also managed relative to the idea statement and the project feasibility—the first control reference point.

Closing

The closing processes are at the end of the project life cycle.

The key point to mention here is that the above presentation is related to the project, not individual stages. In other words, this presentation matches the concepts of the ISO process groups where one can initiate, plan, implement, control, and close the project or phase.

7.5.3 Project and Stage Perspectives

Figure 7.7 combines the project and stage perspectives.

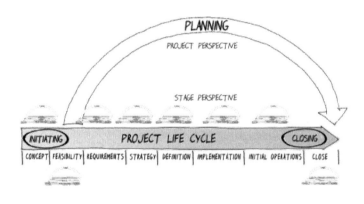

Figure 7.7 The project life cycle and repeating process groups.

The items represented by the smaller images and text reflect the stage view, whereas the large image and text represent the project perspective.

Please note, for clarity:

- The process groups repeat for every stage, even the Close Stage.
- The implementing and controlling arrows for the project perspective are not shown in order not to clutter the graphic.

7.5.4 Can We Combine?

Can a single set of process groups be used?

In other words, can one merge the two perspectives—project and stage—and treat them as one?

In theoretical terms, this merger cannot take place unless there is a single-stage project, which is rare. However, in practice, it is possible to combine them for simple projects.

7.5.5 Can We Consider a Project as a Program?

Some professionals continue to struggle with this concept (the repeated process groups). They ask, "Can one consider the project as a program with multiple projects; each stage being a project?" In other words, they still think of the process groups as phases. Their view (a way to rationalize their understanding) is that to apply the process groups, one must have a project, not a phase (process groups are used only once at the project level).

An easy answer would be, "Sure, why not?" but that would not be correct. To explain, one must go back to the definition of a project (see inset). The *PMBOK® Guide* focuses on the output. However, the author's definition is that a project must produce output (product) and an outcome (capabilities to deliver benefits) to the organization. The benefits must be tangible.

In this regard, let's consider the Engineering Stage of a power-plant project. One can say the engineering design package is output, and we agree. Because it is output (unique result), some may consider the engineering phase as a project; we do not agree.

Here one must ask:

- What is the outcome of the engineering design to the project owner's organization?
- Does engineering on its own deliver benefits to the project owner?

On its own, there is no benefit.

Unless the organization uses the design to construct the power plant – and – the power plant produces power – and – the owner company can sell that power, there are NO benefits.

In other words, the deliverables of the various stages of a project are of limited value on their own and do not directly result in the realization of benefits. There is no value unless these deliverables are all integrated to deliver the ultimate product (of the project) and achieve the expected benefits. In other words, a feasibility study is useful to help a project owner make a decision, but if the team does not deliver the final product, there is no value. The same concept applies to a Requirements Stage, Basic Design Stage, or Engineering Stage, as discussed already. Even construction of the facility is not value added unless the facilities are commissioned and produce power.

Back to the question, can we consider a program with multiple projects?

> **PMBOK® Guide Project Definition**
>
> The *PMBOK® Guide* definition of a project is: "A project is a temporary endeavor to create a unique product, service, or result." (The Project Management Institute 2017).
>
> In other words, the focus is on the output, not the outcome. Is outcome missed, implied, or not part of this definition?

- If one's definition of a project is limited to an output, then yes, because each stage does produce output (although academically and theoretically this is not aligned to ISO 21500, *PMBOK® Guide,* or other definitions).
- However, if one's definition of a project includes the outcome, then a project must consist of all the steps and deliverables to deliver the ultimate product (output). The product provides the capabilities to produce an outcome and realize benefits.

In closing, we cannot consider the project as a program, and we have a project life cycle consisting of stages. Otherwise, every project is a program.

7.5.6 How About Small-Simple Projects?

Chapter 41 elaborates on the concept of project size and complexity via a simplified classification system. However, it is important to stress here a statement that will be repeated more than once for emphasis. *Managing projects requires tailored approaches.* From start to finish, every step has to be appropriate to the project size, importance, and complexity. Although the steps, phases, and stages might be similar, the level of effort would vary widely. Therefore, for small–simple projects, maybe there is no need to repeat all of the process groups for every phase or stage. Maybe the practitioner can do well following the process groups only or a project life cycle only, with the processes spread along the life cycle.

Again, tailoring is a must, and this is why the name of the model includes terms such as *customizable* and *adaptable* to reflect the need for an adaptive mindset.

7.6 Closing Comments

Figure 7.8 includes another representative illustration comparing a generic project life cycle with the process groups and subject groups.

The following is an explanation of this figure.

- The top layer shows a generic project life cycle with four phases.
- The middle layer presents the repeating process groups.
- The bottom row represents a reminder that the processes repeat across the project life cycle.

A reminder: The proposed three-dimensional model is built on integrating the concept of the process groups into a project life cycle. The project life cycle represents the first dimension of the model. The processes and functions represent the second dimension. Therefore, at a minimum, managing projects requires the effective integration of these two concepts as well as integrated management at the project level and the phase level.

Clearing a Critical Challenge 41

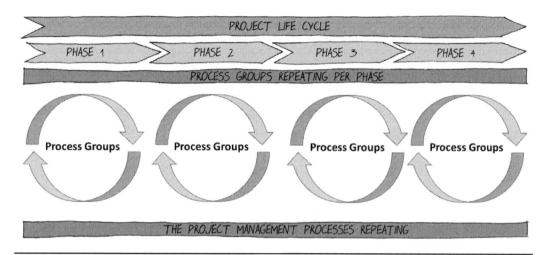

Figure 7.8 Mapping the process groups to a generic project life cycle.

Chapter 8

Various Challenges

8.1 Introduction

What is missing from the *PMBOK® Guide,* ISO 21500, and the other standards? What are some of the inconsistencies and areas not sufficiently emphasized?

It is important to stress that missing items do not indicate *shortcomings* or *errors.* The missing items, at least most of them, are missing *by design,* which means the intention of the original work is to not have them in project management guides. Also, in the context of this chapter, "missing" does not mean there is no mention of these topics at all—they are just not covered in detail for practical use by the project management community.

The following is a list of these topics.

- A methodology
- Organizational system
- Tailoring and customization
- Project classifications
- Templates and forms
- Project life cycle
- Benefits realization

8.2 What Is Missing or Not Covered Enough

8.2.1 A Methodology

As explained earlier, because the IPMA, PMI, and ISO standards are designed to be generic guides, not industry or application-area specific, they do not offer a methodology. The guides advise the readers that they can use other guides or internally developed methodologies to supplement them.

A method or methodology is probably an essential element, along with the organizational system that is necessary for effective project management. Because methods are a function of types of projects, the various associations avoid offering practical guides.

8.2.2 Organizational System

The basis of these different guides is that they cover the processes required to manage a single project (a generic project) or the competence elements that should be acquired by project personnel. The guides' design is based on the assumption that an organizational project management system (OPMS) already exists. This position is understood, and one can agree with this approach and emphasize that these standards should not include an OPMS.

8.2.3 Tailoring and Customization

These guides do not include industry/application-area–specific processes or subject areas. Therefore, it is important for organizations that want to adopt ISO, IPMA, or PMI frameworks to understand that proper project management requires customizing and adapting a model to fit a specific purpose.

The sixth edition of the *PMBOK® Guide* includes a significant focus on tailoring; however, that is different from what we mean here. In this book, the focus is on tailoring methods to be part of the organizational project management system to use on all projects, whereas the guide's focus is on tailoring by the project manager for specific projects.[1]

8.2.4 Project Classification

How to rank projects or classify them in term of size, complexity, or other factors and how to manage per the different classifications are also outside the scope of the various guides.

Project classification is important, because one should treat small projects differently from large projects, and simple projects differently from complex projects. Even within the same organization, project management can vary depending on the project class. Professional associations might have addressed this topic indirectly; however, practical guidelines are missing.

8.2.5 Templates and Forms

The guides are not manuals and cannot and should not be.

Projects are not one-size-fits-all, and the way to manage them is not set in stone. Therefore, standards and guides cannot offer fixed or set templates that would apply to all types of projects. As for a method, these have to be custom fit to the organizational context. Templates and forms should be part of the organizational project management system, as discussed already.

---

[1] A blog post by the author expands on this point. http://blog.sukad.com/20170924/what-does-tailoring-refer-to-in-the-new-pmbok-guide/

8.2.6 Project Life Cycle

Because guides are not industry or domain specific and are not methods, they cannot offer a fixed project life cycle. It is important to note here that in the earlier editions, the *PMBOK® Guide* included a few sample project life cycles from different industries. That was helpful, and PMI should consider reinstating them.

8.2.7 Benefits Realization

The definition of a project in the guides focuses on the output. The *PMBOK® Guide* defines the output as *a product, service, or result* (Project Management Institute 2017), whereas a project success definition should consider the *outcome,* and the outcome is about benefits. Therefore, one must distinguish between output and outcome.

The output is the project's end product, but outcome focuses on result and benefits realization. Consequently, at least from the project owner's perspective, limiting the project success definition to scope, time, cost, and quality is not enough. Project success must include the success of the objectives and realize the expected benefits. One cannot assess this dimension of success at completion or acceptance of the product. Completion and acceptance do not necessarily mean success!

Project success must include the success of the objectives and realizing the expected benefits!

For some associations and in past literature, the topic of benefits realization is a *program*-management topic, rather than a *project*-management one. However, the alternative perspective is that every project must deliver benefits, whether or not it is part of a program. A program also has benefits built up from the benefits of the individual projects, along with other specific benefits that are project specific. Furthermore, benefits realization should be directly related to the definition of what constitutes a project.

PMI has not put much emphasis on this topic until recently. The sixth edition now discusses benefits management and the need for a benefits management plan but considers these as business documents and as pre-project. GPM's PRiSM™ already includes the concept of benefits realization in its guide.

8.3 Other Challenges in Published Guides

There are other challenges in addition to the topics listed already, and a few are listed below. These challenges are related to the published guides and, indirectly, to project performance in organizations following these guides without establishing their own organizational project management systems.

8.3.1 Planning

Half of the processes in the guides are planning related. However, the author believes that there is a weakness. There is a hidden confusion between *project management planning* and *project detailed planning*. Some knowledge areas split the two types; others do not. ISO 21500 mentions the need for support and product processes but does not address them. Maybe this is to maintain the focus on project management, but project teams have to deliver projects and not manage an abstract. Therefore, the more the project team knows, collectively, the better.

How to Improve?

The proposed approach is to consider the interest of a project owner to deliver a project from start to end. The project owner's team needs to move per a given project life cycle, and along this time span there are project-management activities and product-oriented activities, along with the support actions. Again, ISO clearly mentions this but does not offer a solution or suggestions. Consequently, the CAMMP™ approach is to split planning into two process groups, one for *management* planning, and the other for *detailed* planning, both of which include product and support actions.

Planning is the subject of Chapter 15.

8.3.2 Project Change Management

In the guides, there is only one process on project change management, and it is part of the integration area. This process does not provide enough coverage of this critical topic. For example:

- There is no mention of the different types of changes in a project.
- There is no clarity on what the control reference is.
- It is not explicitly clear if there is one baseline or more along the project life cycle.
- There is no clarity on how to fund changes and whether all approved changes will modify the baseline or not.

On capital projects (industrials, utilities, real estate development, etc.), frequent changes can be a significant factor and are likely the leading cause of project failure.[2]

It is also common to observe among project management students and clients (outside the capital project industry) that change is something casual and often can happen without any documentation, review, or prior approval. Maybe that is acceptable in Agile, Scrum, or extreme project management, but it is not for "most projects, most of the time."

Chapter 36 is dedicated to change management.

8.3.3 Project Success

Project success was not a dedicated topic in past editions of the *PMBOK® Guide*. The fifth edition included a brief definition, but it was far from adequate. The sixth edition modifies the definition of project success and links it to benefits management.

---

[1] This is due to the cumulative impact and consequential impact of changes.

Project success varies depending on the context of the project, and it must reflect the position of the organization interested in this question. For example, for a service provider, success could mean profit and customer satisfaction, whereas for a project owner, project success can have multiple dimensions.

SUKAD has developed a four-dimensional model for measuring project success that considers technical success, project management success, project delivery success, and business objective success.

Project Success is covered in Chapter 28.

8.3.4 Pre-Project

Per the guides, the project conceptual–feasibility phase, which covers the time span from idea to decision on authorizing the project, is called *pre-project* and is outside the project life cycle.

Why is this?

Is not the pre-project work (pre-project charter) a phase like any other phase of the project?

Remember, the guides focus on the processes to manage a phase or a project; therefore, by excluding this early phase from the project life cycle, are these guides telling us that the process groups do not apply in this phase? Wouldn't this be inconsistent with the mandate of the guides and successful project delivery?

One might point out that the guides exclude the pre-project phase because the project is not authorized yet. Well, from a project owner's perspective, once management approves the idea/concept and authorizes a feasibility study, the project starts. That is project work that is happening before formal authorization. Keep in mind that in a stage-gate process, the organization can still stop the project after the initial approval and issuing of the charter.

Another might point out that the pre-project phase is excluded because an organization unit other than project management is responsible for it. The sixth edition of the *PMBOK® Guide* clearly refers to the pre-project work as business documents that would be used as input to the project. It is a common practice that an organizational unit outside project management does carry out this work. However, one needs to question whether such an organizational unit is sufficiently aware of the technicalities to be able to determine estimates and timelines for a given project completely. Even if they do, shouldn't they follow the project management processes?

Furthermore, let us look at the other side of the project life cycle, toward the end. At one point in time, the project management team will hand over the product of the project to the end users to start initial operations; before that, there is a need for operational readiness. Should these be excluded from the project life cycle because non-project–management personnel performs them?

> The project work must address all of the work required to deliver the product and realize the benefits, regardless of who does what.

Project management must be alive in all aspects, and the project work must address all of the work required to deliver the product and realize the benefits, regardless of who does what. Therefore, the pre-project work is a phase—a critical phase—and project management skills are vital in this early, ambiguous phase.

8.4 Monitor and Control

The origin (or inspiration, direct or indirect) of the process groups build on the Total Quality Management principle of Plan, Do, Check, Act (the PDCA Cycle), as mentioned in ISO 21500 and as outlined in Chapter 4 (Deming, n.d.).

In comparing the PDCA Cycle with process groups, what can one notice? The guides combined *Check* and *Act* into one process group, either Controlling (ISO) or Monitoring and Controlling (PMI); instead of two separate types of action, Check (Monitoring) and Act (Controlling), now there is one group.

Is this an issue?

Is there justification for splitting monitoring from controlling?

For some projects, it is not a big deal, and combining the two actions is satisfactory. However, at least in capital projects, monitoring is usually a role for the project management team members who monitor the performance (procurement, cost, progress, quality) and compare actual results versus plan. The team member activities may include performance analysis and recommendations for actions (corrective or preventive). Conversely, controlling is an action for a project manager or sponsor, and it may often require management decisions through approved actions or changes.

In closing, it is a preference to split monitoring from controlling, in line with the original PDCA concept. However, this point may require a task force to consider, and the author's recommendation is not to split at this time.

8.5 Conclusion and Recommendations

How can one decide if these items are valid?

Are there other inconsistencies or missing items?

These and similar questions are necessary to improve the practice of project management and to close the gaps caused by these challenges.

As it is time to end this chapter and Section I, it is vital to remind the readers that Section I, in its two parts, were written to address the current state of project management and the challenges in practice, especially from limiting the practice to a single guide from a single professional association. It is critical to look *into* the box, since many of these guides do offer the community valuable resources, but it essential to consider what is *outside* the box (or to look into the various boxes).

It is time to shift focus to the solution, starting with the organizational and strategic aspects and rationale for the universal, three-dimensional methodology, then a part dedicated for each of the dimensions.

Section II, Part C
Overview of the Offered Solution

Chapter 9

Organizational Project Management

9.1 What Should Organizations Use?

The first step in discussion of any solution to the project management challenges in the current practices is to start with an organizational perspective. Reflecting on Chapter 1, what should organizations use to manage projects? A method, a framework, a guide? What would be appropriate?

Leading practices would be for organizations to develop their *organizational project management system* (OPMS). Such a system, in the context of this book, is about the various components and elements of managing projects. It might be necessary to widen the spectrum and incorporate the management of programs and portfolios. However, keeping the focus on projects for now, the OPMS must have an organizational focus rather than a single-project focus.

To reinforce the point: for managing the organization's projects, it is vital to establish a sustainable organizational project management system.

> **OPMS**
>
> An organizational project management system is about the various elements and components for managing projects within an organization. This system includes policies, processes, methods, procedures, guidelines, templates, and flowcharts.

9.2 OPM and Professional Associations

What should be the knowledge reference for the OPMS? Should organizations follow PMI, IPMA, GPM, or something else?

As outlined in earlier chapters, organizations can follow any one of these associations and *build on* what they offer. However, organizations must supplement what these associations offer to close the gaps.

Let us elaborate on the gaps (a refresher).

If one truly studies and understands all aspects of these associations and what each one of them does or doesn't offer, then the product of one association, with supplements, might be enough.

- PMI's and ISO's strength is *project management processes*.
- IPMA and its member associations are known for *competence baselines*.
- PRiSM™ is a method, but its strength is on *sustainability*.

What the author and SUKAD have done is to simplify what exists as much as possible and integrate the learning from all of the above into an organizational system, inclusive of a methodological approach.

9.3 What Are the Components of an OPMS?

One standard or method is not enough to manage an organization's projects effectively. OPM must adopt a system thinking approach, considering all aspects, including the learning of the various project management associations—the ones with a general project management perspective and the ones with a focus on areas of specialties (subjects) within project management. If an organization does not want to become familiar with all of these, they have one of four choices:

1. Do nothing and continue to manage projects by common sense, without a system. However, these organizations must realize that it is common sense not to depend on common sense alone to manage projects. An exception might be in micro or routine small projects.
2. Start to build their internal OPMS from scratch.
3. Start with something like what this book offers, or similar products in the market.
4. Follow one of these associations, using a consultant who can decipher some of their mysteries.

Figure 9.1 is a product of the SUKAD Research and Development work, and it represents a project management maturity model, which is also a framework for building and sustaining the OPMS. The model consists of three sets of elements: the *fundamental elements* (the three elements in the core), the *differentiating elements* (three circles), and the *strategic element* (the outer circle).

The fundamental elements of this OPM System are:

- **1D, Project Life Cycle.** This element refers to the first dimension of the CAMMP™ Model, a project life cycle. One can use the CAMMP™ standard project life cycle, a modified project life cycle, or an internally developed approach.
  Section II, Part E addresses the CAMMP™ standard project life cycle.

- **2D, Processes and Functions.** This is the second dimension, with the process groups, processes, and functions from ISO 21500 and the *PMBOK® Guide*, but as modified by CAMMP™. For this topic, the various specialized associations offer organizations useful resources. Just a reminder, some of these associations would be AACE, SAVE, Guild of Project Control, and others.
  Section II, Part D addresses the various processes and functions.

Organizational Project Management   53

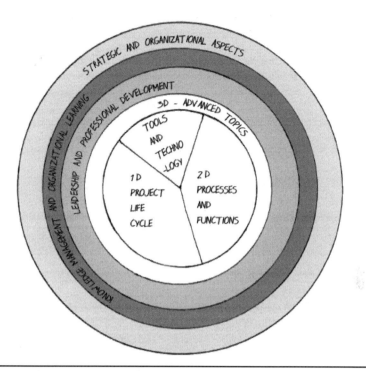

Figure 9.1   The Seven Elements of Project Management Maturity™.

- **Tools and Technology.** This is not unique to CAMMP™; rather, it covers any software or specific tools that might exist or be internally developed.
  This book does not address tools and technology.

The term *fundamental elements* emphasizes that these elements are the minimum required to manage projects and achieve success. On the other hand, the term *differentiating elements* represents the elements that are essential for a higher level of performance and project management maturity.

The differentiating elements (the circles) are:

- **3D, Advanced Topics.** These are topics listed as the third dimension of CAMMP™ and deal with subjects such as competence, sustainability, best practices, and project success, which the author believes to be a must for advancing the field of project management.
  Section II, Part F addresses the third dimension.

- **Leadership and Professional Development.** For this element, there is no formal association listed here. There are numerous resources and books on project management, leadership, and professional development.
  This book does not address this topic directly, but Competence is a related topic.

- **Knowledge Management and Organizational Learning.** This element covers such things as the corporate knowledge base, lessons learned, tools and approaches for sharing knowledge, among other actions and components.
  This book does not address this topic directly, but it touches on it in various chapters.

The strategic element (the outer circle) is:

- **Strategic and Organizational Aspects.** This element is about the proper organizational aspects such as project governance, portfolio management, and strategic planning.

> The big picture is the organizational project management system (OPMS).

9.4 Closing Comments

The repeated hypothesis here is that a standard document, a guide, a set of processes, or a method—taken independently—are not enough to manage projects effectively. *Once again, organizations need a system approach.*

Consequently, the proposed model—the advocated solution—is to provide an integrated approach from the various resources available within the project management domain. Organizations can either develop their OPMS by tailoring the SUKAD approach or kick start their effort using any other suitable approach.

The big picture is the organizational project management system. The rest of the book is related to the methodology.

Chapter 10

The Three-Dimensional Model

10.1 Introduction

It is time to introduce the three-dimensional model.

The previous chapter addressed the need for an organizational project management system (OPMS), which is essential. Two of the core elements are project life cycle and processes and functions. Building on this concept, a valid question would be how to build a universal project management methodological approach as a core component of an OPMS?

Let's expand on the question: How do we build a universal methodological approach for managing projects that is practical and flexible enough to adjust for project type, domain, classification, or a learning platform?[1]

First, it is vital to provide definitions of the relevant terms.

10.2 A Few Definitions

10.2.1 Universal

Universal DOES NOT MEAN one-size-fits-all.

The intent for the use of the word *universal* is that the approach is not limited to one global association, such as PMI, IPMA, ISO, GPM, or others. Rather, it uses elements from all to come up with a robust approach that is scalable to project classification, tailored to an organization's environment, and expandable to consider more than the core elements of project life cycle and processes.

---

[1] In this context, platform refers to the international standards by project management associations.

10.2.2 Method or Methodology

Chapter 1 presented the differences between method and methodology. Based on the definitions from that chapter, the author recognizes that for a given situation—a particular project context—a project team must follow a method. However, in the context of this book, the emphasis is on a management approach that is not restrictive or limited to a given project context. This approach is universal, customizable, and adaptable, *a system of methods used in a particular area of study*, so *methodology* is the appropriate term in this context.

A universal methodology (methodological approach) is one that, once customized and adapted, will produce a set of methods, each for a particular project type, domain, or classification.

10.2.3 Project Type

A project type, in this context, refers to a kind of work that the project aims to deliver. Project types include things such as marketing, financial, training, or other types of projects.

10.2.4 Project Domain

A project domain refers to an industry such as technology, telecom, real estate, petroleum, software development, etc.

10.2.5 Project Classification

Within a particular type and domain, one could also classify projects based on size, complexity, and other such factors.

10.2.6 Platform

Platform refers to international associations and their approaches to managing projects. Repeating some of the points from earlier chapters:

- PMI's *PMBOK® Guide* and ISO 21500 focus on processes and functions (knowledge areas/subjects).
- PRINCE2® offers a method comparable to what we present here, but CAMMP™ is more flexible and wider in scope.
- IPMA offers Individual Competence Baseline® and now Organizational Competence Baseline®, so their focus is competence.
- GAPPS, an organization that does not offer certifications, also concentrates on the competence of specific roles in projects.

- GPM® Global's strength is on sustainability and has a method for integrating sustainability into project management.

Other organizations offer narrower scope, such as AACE® International, which focuses on cost and planning; SAVE International, which focuses on value engineering; the Guild of Project Controls; and others.

10.3 Principles of Building a Method(ology)

Building on these definitions, how can organizations build such a methodology?

What are the principles?

A proper method (methodological approach) must incorporate the following fundamental principles:

- A project life cycle that can be tailored (customized, adapted) to the project type, domain, and classification.
- A set of project management subjects and processes to apply to the phases and stages of the project life cycle.
- Deliverables and gates that would be applicable in every stage and phase, and along the project life cycle.
- It must consider the various project processes: project management processes, product processes, and support processes—in other words, it expands the scope of management of projects to the delivery of projects.

The above are the principles for the foundation and will result in a fundamental methodological approach, but is not enough in search of excellence. To build on the above and ensure that organizations have a truly universal approach, they need to:

- Consider project classification; one-size-fits-all DOES NOT work.
- Incorporate other elements such as competence, sustainability, leadership, innovation, pioneering mindset, and other factors that would be vital for excellence.

The CAMMP™ Model offers such an approach, and it was developed to fulfill these principles.

10.4 The SUKAD CAMMP™ Model

The SUKAD CAMMP™ Model (The Customizable and Adaptable Methodology for Managing Projects™) is a universal, three-dimensional model (3D). We have covered the basic principles above.[2] In this chapter, the focus is on the high-level aspects; Chapter 13 elaborates further, and the rest of the book extensively covers the details of the approach and its application.

---

[2] To be transparent, our earlier work on CAMMP™, versions 1 and 2, were mostly focused on the first two dimensions. The third dimension has been evolving from working with CAMMP™ for almost 10 years.

10.4.1 The First Dimension

The first dimension[3] is the project life cycle (the horizontal dimension), which helps practitioners follow a project from idea to closure (and beyond). The project life cycle consists of phases, stages, deliverables, and gates. What CAMMP™ offers is a standard model that can be tailored. (Tailoring is covered in the last part of this book.) The intent of assigning "the first dimension" to the project life cycle is to emphasize that a project life cycle is the most basic form of a methodology and must be the starting point.

10.4.2 The Second Dimension

The second dimension[4] is the vertical dimension of the project life cycle. *It is about the application of the project management processes and process groups along the project life cycle in every stage or phase.* The CAMMP™ Model offers a modified version of the process groups from what ISO and PMI offers.

10.4.3 The Third Dimension

The third dimension[5] consists of layers, layers that organizations could implement along (and on top of) the project life cycle and the processes. These layers are what help organizations move from a fundamental methodological approach and transform it into a highly sophisticated and robust system that will enable and empower them to seek excellence.

In other words, the basic approach (two dimensions) could be of value for organizations starting on their project management journey, and as they become comfortable with the approach, they can implement the third dimension, one layer at a time.

10.5 Methodology Main Characteristics

Figure 10.1 is a view of the standard model, but only the first dimension—the project life cycle with phases, stages, and stage gates. As stated earlier, Chapter 13 will offer a brief elaboration on the model and the image provided here.

10.6 What Are the Key Features of CAMMP™?

1. CAMMP™ is a methodology for *managing (delivering) individual projects.*
2. The methodology takes a holistic view of a project; it is a *project life-cycle* approach.
   a. The methodology *emphasizes the various components of projects such as phases, stages, and stage gates,* in addition to functions such as scope, time, or risk.
   b. The project life cycle covers the whole project, *starting with the idea* for the project and ending with project closure, including a focus on processes within a stage.

---

[3] Discussed in detail in Part E.
[4] Discussed in detail in Part D.
[5] Discussed in detail in Part F.

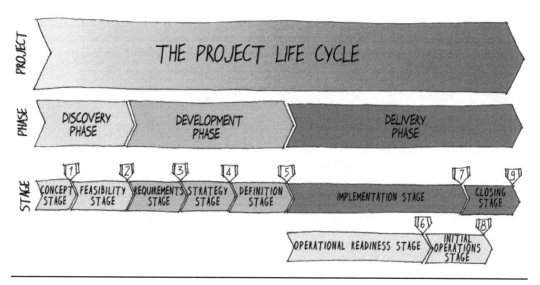

Figure 10.1 The Customizable and Adaptable Methodology for Managing Projects™ (CAMMP™).

3. The intention is for the methodology to be *simplified, practical, and systematic*. However, "simplified" does not mean "simple."
4. It emphasizes a *disciplined approach* to project management, balanced against the need for *flexibility*.
   a. Discipline is achieved via the distinct stages and stage gates.
   b. Flexibility is achieved via the customizable and adaptable principles.
5. The gates (at least some of them) must be formal interim reviews that require acceptance of prior stage work and authorization to proceed to the next stage.
6. Although the methodology includes a standard model, one size *does not* fit all, hence the name *Customizable* and *Adaptable* Methodology for Managing Projects™.
7. The methodology aligns to popular global guides and project management frameworks, such as ISO, PMI, and IPMA, but these frameworks do not constrain it.

In summary, it is worth noting that the methodology described here is a common approach, and many global organizations with mature project management systems use one form or another of the concepts. In other words, *it is proven practice*.

10.7 What Is Not New

The concept of a project life cycle is an established and known concept.

The gate concept is also well established. Many global organizations use a gated process.

Many organizations have developed similar methodologies as part of their project management system. In general, methodologies are custom fit to organizational requirements. This tailoring and proprietary nature are the main reasons few project management methodologies are published and widely known.

10.8 What Is New (or Not Well Known)

What is new and unique? What has not been emphasized in common practice and is stressed here?

1. **Across industries.** As stated earlier, the principles of this methodology have been in practice in mature project management industries for a long time. However, most of the organizations in those industries do not publish their internal processes, and any published information likely has limited circulation. Therefore, what is new is an attempt to transfer, in a simplified and practical way, the expertise of mature project management industries across to other industries that are newer to project management.
2. **Gates.** Gates are not a new concept; what is new is to stress the gates as critical factors for enhancing project management performance across the entire life span. Furthermore, a significant number of organizations do not have any formal stage-gate process.
3. **Principles of customizing and adapting.** The third and most fundamental concept is that the methodology is fit-for-purpose instead of being rigid and therefore making it difficult to apply one model to all situations, domains, or project classifications. There are universal principles in project management that cross industries, yet there are applications and approaches that are unique to an industry or a project class that one cannot ignore.
4. **Strategic aspects.** There are four key examples of strategic aspects that are part of CAMMP™:
    a. Incorporation of the project's first stage as an essential component of the project life cycle; what PMI labels as *pre-project*.
    b. Introduction of a gate focusing on the strategic alignment of the project to organizational objectives—the first gate.
    c. Inclusion of project close as part of project delivery, instead of a separate stage that can be ignored.
    d. The four dimensions of project success and the link of at least one of the dimensions to benefits realization. In other words, for project owners' organizations, project completion and closure is not the end.
5. **Project close as part of project delivery.** This strategic aspect merits further elaboration. Some organizations do not put enough emphasis on properly closing a project. However, by making project close a defined stage within the project delivery phase, and with a gate required for the exit, project close will be difficult for organizations and their project managers to ignore.
6. **Moving control baseline (control reference points).** This is not a new concept but is poorly understood by many practitioners. The main point is the discussion of project control and baselines. Most project management literature, including the *PMBOK® Guide*, states and emphasizes that control is against the plan—which plan, it is often not clear. Is it the project management plan? Chapters 35 and 36 include coverage of this topic.
7. **Operational readiness.** Since project management personnel often do not carry out operational readiness activities, they are easily overlooked, or the potential for full integration with the main implementation activities of the project is lost. Operational readiness and initial operations have their dedicated stages in the project life cycle and are covered in Chapters 24 and 25.
8. **Dimensions of project success.** This topic was also mentioned earlier and is the subject of a dedicated chapter, Chapter 28.

10.9 Agile/Scrum/Waterfall

There has not been any significant mention of alternative approaches such as Agile, Scrum, iterative, and incremental in this book so far. The main reason for this is because *the CAMMP™ Model is an adaptive model,* which can incorporate Agile, Scrum, or incremental principles. The author does not believe in Agile or Scrum as standalone approaches or methods for managing a project end-to-end from a project owner's perspective.

Here is why!

All projects, from a project (business) owner's perspective, must start with an idea, alignment to strategic objectives, feasibility, and authorization. These steps are in line with the approach presented here: following a traditional life-cycle model with phases and stages. In other words, project owners have to follow a phase-stage approach. It is also worth noting that often the early stage(s) are led by the business owner (internal or external).

After a certain point along the project life cycle, organizations may assign the project to developers, whether internal or external service providers—let's call them the *development agency*. In this case, the development agency is working on a single stage (or combining two or three stages) of the business project life cycle. For this stage, or more, the development agency may use Agile principles. Therefore, the agency may say they are using Agile methodology, but in reality, they are working on a piece of a project that has phases and stages, and Agile or Scrum usage is limited to a part of the project—one or more phases, not the whole life cycle.

Could Agile be used start to finish? Maybe, but how?

To summarize (and partially repeat), what is different in the Agile or incremental environment is that a project may start by following a waterfall-like approach, but when the project reaches the stages of definition and implementation (requirements, design, development, testing), the project team can merge these stages to follow the Agile or incremental approach. In other words, one or two phases of the project could adopt Agile or Scrum, but from within CAMMP™. In the last few years, it has become apparent that "WAgile" or "AgiWater," whatever name one uses for the blended approach, is becoming popular due to the shortcomings of the pure approaches.

In closing, the CAMMP™ Model is not a Waterfall model, it is an adaptive model. The stages in CAMMP™ may overlap, and the degree of overlap is a function of the project and organizational preferences. The author's view is that any project must follow a project life cycle, with the early phases done in a traditional, sequential approach (with some overlaps). Then, depending on the nature of the project, it would either continue to follow a sequence of the stages, building on each other, or, once the project reaches a point at which Agile principles will add value, the project owner or development agency can adopt them, then finish the project with a proper close stage.

Figure 10.2 presents a possible model of how CAMMP™ can integrate Agile or Scrum principles into its project life cycle. Two points should be raised here in relation to the image.

1. The Requirements Stage is to reflect the business requirements, may be high level, and would be different from the Requirements element in the circle.
2. Although the figure shows one cycle, the intent is that this cycle would repeat based on the project's needs, such as the segments durations or the size of the work increments.

62 Project Management beyond Waterfall and Agile

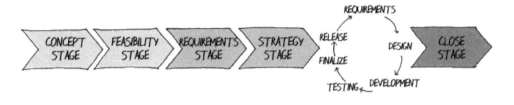

Figure 10.2 Possible project life cycle model integrating adaptive principles.

10.10 Closing Comments

This three-dimensional model offers to integrate the learning from leading organizations such as GPM Global, IPMA, and PMI, along with the author's own research and development work, to build a truly universal model—a methodological approach to managing projects. The standard model will be generic, but it would be tailored to fit the desired project domain, type, and classification.

Chapter 11

Rationale and Model Perspectives

11.1 The Story That Led to This Work

The author was leading a class attended by about 10 professionals. They were all from the same company but different functions, such as public relations, human resources, marketing, call center, finance, strategic planning, and an engineer working on a PhD. The class was an overview of project management—an introductory workshop.

During the class, the public relations person could not even comprehend the concept of budget or estimating because "that is something handled by an account manager"; she wanted us to explain to her the project life cycle that reflected her function and her practice.

The PhD candidate wanted a model to help him work on his thesis.

The human resource person and a couple of other participants did not appreciate that the class included discussions on scheduling. They did not want a discussion on scheduling, although, and almost without exception, all had complained about unrealistic schedules, delays from the different stakeholders, and failures to deliver on time.

In short, the key issues were that project management seemed too complex for them, and they wanted case studies and sample projects for each of their disciplines. They felt that the only way to learn project management is through samples and examples that directly related to their work. In other words, the *PMBOK® Guide* and its 40+ processes were not something they could readily relate to.

On the evening of the course, the author and two colleagues, both of whom are international consultants in project management with years of experience, vented their frustrations on the lack of published practical examples that relate to a methodological approach. That evening led to a decision to do something about it and started to formalize the idea for a new methodological approach. That night witnessed the birth of The Customizable and Adaptable Methodology for Managing Projects™ and the SUKAD research and development program. That experience was one of the driving factors for a series of books and e-books, all leading to this work.

11.2 The Challenges for Professionals

The biggest challenges faced by professionals, like the group in the story above, are:

1. They want to use a model that is relatively painless, or at least simplified, over what exists in the market. In other words, not everyone wants to be a professional project manager or acquire professional certification in project management.
2. While simplification is a crucial criterion, they do not want something so trivial that it is not effective.
3. They want samples that deal with *real projects* and not tasks disguised as projects.
4. They want a model that guides them in the various steps of the *project*.
5. They want a model that includes controls and not a continuous run through the *project* without checks and balances.
6. They want an approach that encourages them to *use* project management instead of *fearing* it.
7. Most importantly, they want to see *project* samples that apply to their work environment.

Concisely, they want something that is customizable and adaptable to their real lives.

We expect that our readers can relate to the above.

11.3 The Birth of CAMMP™

It was time to convert theory into action, and the Customizable and Adaptable Methodology for Managing Projects™ (CAMMP™) is the answer. The primary success factor for this model is its vitality to deal with the entire set of challenges, resulting in a comprehensive, simplified, practical, and straightforward methodological approach for the professional community. Its main features must include:

- Suitability for both novices and experts
- Customization toward industries or organizations
- Adaptability for specific functions or project classifications

A comment from a reader of a blog post on the model asked, "Shouldn't all methodologies be flexible, so one can customize them?" Agreed, but this concept is not obvious for most practitioners.

11.4 Full Project Life Cycle Perspective

By emphasizing the word *project* in the numbered list in Section 11.2, the intent is to stress that most of this work, unless otherwise noted, takes the perspective of the entire project life cycle, not a single phase or stage, process, or function. Understanding the difference between stage work and project life cycle is a vital distinction, because many practitioners may only have been exposed to a *stage* of a project. For a service provider, the whole project may be only a stage from their client's perspective. For example, for a construction contractor, projects are to construct facilities. On the other hand, for the facility owner, construction is only a stage preceded by engineering design, which is, in turn, preceded by concept design and feasibility study.

In keeping with the above, note that the methodology presented in this book reflects a project life cycle that ranges from the idea to project closure, providing the reader the full spectrum of a project's life.

11.5 Multiple Stakeholders

A common definition of a project's stakeholders is *everyone who is involved or impacted by the project*. Typically, *involved* refers to those who have a role in the delivery of a project, which would include the project teams. *Impacted* relates to those who will be affected, positively or negatively, by the outcome of the project. One can expand this definition to include those who will influence the project for one reason or another.

On most projects, the stakeholders will come from a variety of sources, internal and external. Therefore, it is common that most stakeholders do not have exposure to the whole project or the full span of the project life.

Consequently, project management must be about *how to manage a project effectively, regardless of who does what*. Maybe one should use to *deliver a project* instead, since it is more encompassing than *manage a project*. A project management model should be about delivering the product, not only managing it. It is necessary to understand the whole picture of a project, regardless of who does what work. The concept of "regardless of who does what" is important to comprehend by professionals whose project management experience is limited to a Project Stage or function.

11.6 Idea Developer (Project Owner) Perspective

Further to the previous point, with stakeholders coming from different sources, whose perspective does the approach consider?

The methodological approach reflects the foundation of a project life cycle, end to end, typically, but not necessarily, from the project owner's perspective. In this context, the reference to a project owner is not particular to a person; rather, it is the organization that is developing the project and will own its product. However, it is important to note that, although the general approach follows a project owner's perspective, the methodological approach will benefit all, including service providers. Chapter 42 addresses how service providers can gain competitive advantage from this model.

11.7 Closing Comments

A point that this book stresses often is that a project life cycle, a set of processes, or a set of competence elements are not enough to manage projects. It is critical to open the boxes provided by the professional associations, gather their valuable content, extract the core principles, and link them through a practical approach. CAMMP™ does not recreate the wheel—it improves it. CAMMP™ builds on what exists to make it easier for all types of professionals to use it. In a way, CAMMP™ offers the frame of a picture puzzle, and the rest of the picture comes from the associations' guides. Therefore, to complete the picture, we need all of the pieces.

Toward the end of the book, we will present ways to think of different projects. We will show samples and simulations of small and simple projects, often managed by accidental project managers, along with large and complex projects requiring years of intensive project management expertise.

CAMMP™ is a methodological approach that blends well with ISO 21500, the *PMBOK® Guide,* IPMA's Individual Competence Baseline (ICB®) and Organizational Competence Baseline (OCB®), the GPM P5™ Standard, and PRiSM™.

Chapter 12

A Vital Concept: The Stage Gate Process

12.1 Project Management Is a Disciplined Approach

Current practice may range from rigid bureaucracy to *laissez-faire* absence of processes, and neither approach is effective for managing projects. Project management must be a *disciplined approach*—one which incorporates a balanced set of decisions and control points along the project life cycle and within the project's phases. These control points are:

- Ensure alignment of the project business case to the organizational strategic direction.
- Determine that the project is feasible and a priority.
- Ensure alignment among executives, line management, the project team, and the various stakeholders before expending many resources.
- Include checkpoints to validate that the project continues to progress in line with the stated objectives and will be successful.
- Lead to the proper delivery of the project's product and formal closure.
- Measure success and the realization of the expected benefits.

In the absence of such control points, how can management know that its projects are moving in the right direction and remain aligned throughout their life cycles?

Another value of these control points is the logical and proper sequence for project development and delivery. A proper sequence, with required approval and oversight, will allow the team to move from one Project Stage to another with the correct amount of detail and information—no more, no less.

In this book, we will use the term *gates* to refer to these control points. The reader will see two types of gates: *stage gates* and *process gates*.

12.2 Definition of Gates

Stage gates are the gates along the project life cycle that separate the stages. These are like phase or stage exits. The author prefers the term *gate* instead of *exit,* because exit means to leave something, so the criteria at these exits will allow people to exit but do not address what comes next. On the other hand, a gate could refer to an exit *or* entry, and in the context of CAMMP™, it is both. Most stage gates have two purposes—one to confirm that the work of the preceding stage is completed appropriately and satisfactorily (exit a stage), then to decide whether to proceed or not (enter a stage).

Process gates are related to the process groups and repeat at every stage.

Stage gates are covered in more detail in Part E, process gates in Part D.

12.3 Project Life Cycle and Stage Gates

In Section 12.1, the topic of proper sequence came up. What is the appropriate sequence of control points and other project events? The sequence comprises all the control points and events of the project life cycle. The project life cycle consists of various phases and stages, and the control points are the review events called stage gates.

How many stages, or stage gates, are there?

It depends!

It depends because the number of gates is a function of the project size, complexity, industry, and other factors such as organizational maturity and culture. There are organizations with as few as three gates and others with as many as twelve. Most have some number in between.

Which is better, to have a few or many gates?

Do more gates lead to increased bureaucracy because of too much control?

Not necessarily!

Having gates for effective project management is crucial. It is acceptable to combine gates, thereby reducing the count, but the *function* of each gate must remain. Management philosophy drives the primary decision factor. If executive management has a high threshold for risk, runs a highly empowered organization, and has sufficient maturity to accept a greater degree of uncertainty, then reducing and consolidating the number of gates can be a consideration. On the other hand, if organizational risk tolerance is low and uncertainty must be minimized, then more control points (gates) are necessary.

A fundamental concept in project management is proper control. How to define proper control is something that cannot be easily explained and is a function of experience and competence.

12.4 Importance of the Stage Gate Process

Within this methodological approach, stage gates are critical events throughout the project life cycle. Important factors for the gates are:

1. With the exception of software projects using Agile principles, experienced project management practitioners will agree that no one should want to implement a project before

completing a detailed plan. Detailed planning should not start before there is an alignment among the stakeholders on the requirements. Furthermore, before requirements, one must have an approved concept; and further back in the chain, there has to be a determination that the project is feasible and that there is a business case.

If there are no gates, then what prevents the team moving from one stage to another without properly completing the prior stage work? Even worse, what would prevent skipping a stage or two if there are no control points? Finally, what ensures that the team is implementing the right work?

2. The stage gates also play an essential role in organizational and project governance. The importance of such gates is especially pertinent as more and more projects are substantial investments; therefore, project governance becomes a keystone of corporate governance, and the gates are essential for successful governance.
3. Another important factor is determining the right level of project investment for each stage of the project. In other words, is the right amount of effort in place to balance the effectiveness and efficiency of the application of project management? In each stage, it is important to do just the required work that is necessary to meet the organizational requirements for the scope of the stage—no more, no less. What is needed is enough progress to make a proper decision on whether the organization should allow the project to go to the next stage or not. Spending more time and money than necessary on a stage could be counterproductive and must be avoided.

You can combine but cannot eliminate stage gates.

Before closing this part, it is important to stress that software development projects are special types of projects and may require a somewhat different approach. However, even for these projects, gates would be necessary, if not at the stage level, then within the cycle for each incremental feature.

12.5 Governance

This book does not include a dedicated chapter about governance, because project governance is a fundamental component of the Strategic and Organizational Aspects element of the SUKAD Seven Elements of Project Management Maturity™. In other words, project governance is part of the organizational project management system. However, methods have to comply with the organizational policies, system, and corporate governance. Therefore, governance aspects are embedded throughout CAMMP™ and in this book.

The previous parts of this chapter emphasized the need for proper stage-gate process and touched on process gates. Gates are a manifestation of project governance. Including strategic alignment at Stage Gate 1, other examples of incorporating proper governance in CAMMP™ are (1) a proper feasibility study pre-project authorization, and (2) incorporating the Close Stage in the Delivery Phase. Further topics that are part of CAMMP™ and that directly (or indirectly) comply with project governance are the chapters in Section III, Part G, such as project approvals, estimating, control reference points, and change management.

12.6 Is Not the Stage Gate Process . . .

12.6.1 . . . A Form of Risk Aversion?

Consequently, one might ask, "Is the stage gate process just a reflection of risk aversion?"

No!

The fundamental point is that the stage-gate process is necessary to ensure that the team is progressing toward the goals established for them, and thus there is a need to check periodically for direction. Like the team, we move toward the goal. The team could be risk averse and follow a safe approach, or the team could be risk neutral or even adopt a "risky" attitude. Therefore, risk tolerance and attitude is an informed choice, and the team must be aware of the issues and risks that could face them. Having said that, one can say that the stage-gate process helps the team to manage risks; there is a difference between risk *management* and risk *aversion*.

12.6.2 . . . A Form of Bureaucratic Process?

One of the main factors in the stage-gate process, regardless of the number of gates, is the requirement that executive management expedite the review process and have the team pass through the gates as efficiently as possible. If this does not happen, the unfortunate consequence is the trap of bureaucracy, which is essential to avoid.

However, this possible bureaucracy is not due to the gate process, but rather to an inherent organizational mindset. In bureaucratic organizations, even a two-gate process can be cumbersome and disruptive to the project, especially when decisions take weeks or even months.

12.6.3 . . . A Killer of Innovation?

Occasionally we hear that the gate concept, a methodology, and even project management in general are "killers of innovation." The theme is that a disciplined approach is anti-innovation because of the imposed control and standardized processes.

We cannot agree!

One aspect of the definition of a project is that it is unique. This uniqueness opens for the team many opportunities for innovation, in all facets of project management. There is room for innovation in planning, in management, in procurement, and in many other areas. The concepts of integrated teams, value-improving practices, zero change or no incidents are examples of innovation in project management. In every industry and every occupation, there are bureaucratic processes, routines, and even semi-robotic actions. On the other hand, there are opportunities for creative thinking, innovation, and excellence, without losing essential controls.

Now with standardized processes, the author thinks innovation can thrive.

Why?

Because the team does not have to waste time and energy reinventing the wheel and trying to figure out what to do at the gates, or what would be a proper template for a given deliverable. Some of this freed-up time can be used to find creative and innovative solutions for the project's challenges.

12.6.4 Case Study

The following is a case study from a mega project that shows a high tolerance for risk using innovative approaches.

In his early career, the author was working on a mega project: developing and constructing a large-scale petrochemical plant on an island. The existing island was not large enough for the plant's space needs, so there was a necessity to expand the island through reclamation. Another challenge was, with the project being on a small island, there was not enough space for laborers to work and house, compounded by insufficient qualified laborers and technicians in the area.[1]

Therefore, to deal with these three greatest challenges and their associated risks, the solutions considered were:

1. Expand the island more than necessary to accommodate a large number of people to work effectively. This was an expensive solution, and it did not solve the challenge of finding the required resources, housing them, and ferrying them to the island.
2. Utilize pre-assembly and modular work—meaning, do as much of the work as possible in fabrication yards or factories in different global locations but primarily in nearby countries. This approach allows high efficiency of the workforce, reduces reclamation for expanding the island, but adds shipping cost and risks.
3. The third option was the author's preference (it sounded unique). This option was to build most of the plant on large ocean-going barges, transport them to the island, anchor them, build an embankment around them, drain the water, fill the area around the barges with compacted rocks and soil, and fill the barges with concrete. This action would transform the barges into the foundations for the plant's equipment, and, indirectly, lead to the reclamation of the necessary additional land.

Option 2 was selected, which resulted in about 200 modules, each the size of a small three- or four-story building.

> **Innovation versus Control**
>
> Once I was at dinner with a colleague, a program manager in the pharmaceutical industry.
>
> I asked: "I'm curious, how do you manage and control research projects in an industry where research and innovation are crucial?"
>
> He responded: "Mounir, even in our industry, management and control are welcome concepts. We do allocate some percent of our research budget toward open research where we do not impose much control. However, for most other research, there are 'control points,' and even the researchers welcome that."

---

[1] Projects like this requires hundreds of engineers and professional staff, in addition to thousands of labors and technicians.

The result: The project finished within the required time, slightly below budget, and with an outstanding safety record. This action was not the only reason the project was successful—there were numerous other innovative and leading practices utilized on the project.[2]

12.7 Closing Comments

The challenge is to differentiate between essential and necessary control versus excessive control; once again, project management is art and science. Project management has to be adaptive and innovative rather than restrictive and rigid.

Project management has to be a facilitating process and not a controlling process! These are some of the fundamental concepts used to develop and update The Customizable and Adaptable Methodology for Managing Projects™.

---

[2] The author has published some of the other innovative techniques as case studies on the SUKAD Project Management Knowledge Portal at http://knowledge.sukad.com/project-management-case-studies

Chapter 13

Brief Overview of CAMMP™

13.1 Introduction

The previous chapters covered the conceptual introduction of a universal methodological approach, the rationale behind it, and the fundamental concepts of the stage-gate process, governance, and other topics. It is time to shift focus and introduce the Customizable and Adaptable Methodology for Managing Projects™, CAMMP™.

Understanding this chapter is critical, as it is the foundation for the following parts.

13.2 History of CAMMP™

The initial version of CAMMP™ came into being in late 2007 to early 2008. It started with workshops delivered to clients, then a book, *The Inheritance,* written using the story-telling format, which was published in February 2010. The work on the model continued with minor updates until the time the second book, *Redefining the Basics of Project Management,* was published, in 2013. This book included the fundamental concepts for explaining the methodology per Version 2. Many of the chapters of that book are now in this book, as it is time to update the methodological approach to the third generation.

CAMMP™ Version 3, presented here, maintains the main concepts and principles used in building the methodology. However, there are three substantial changes from the previous version:

1. The approach changed from a two-dimensional model to a three-dimensional model. Some of the components of the third dimension were there before, but were not clearly split out as an independent dimension.

2. The standard project life cycle model is modified. It maintains the three phases, as we will see later, but it changed the name of the first phase. It also added three stages for further clarity.
3. With this change, it was necessary to address how to manage projects from different classifications. A later chapter shows how to manage projects using three simplified categories: *small-simple projects, medium–moderate-complexity projects,* and *large-complex projects.*

13.3 The Model's Brief Explanation

13.3.1 Overview

It has been established that the model is built on a *project life cycle model,* which starts at the idea and finishes with project closure. The main components of the life cycle—the first dimension—are phases, stages, stage deliverables, and stage gates.

Notice that CAMMP™ uses both terms, phases *and* stages, whereas many practitioners consider them interchangeable.

13.3.2 Phases

The term *phase* refers to a period of the project life cycle with a specific purpose and a significant output. The standard model has three phases, specifically, *Discovery Phase* (previously Concept Phase), *Development Phase,* and *Delivery Phase,* per Figure 13.1. These phases should be valid for any project, regardless of size, complexity, domain, or type. They are fixed, and although some organizations may choose different terms for the name of the phases, they need these three phases.

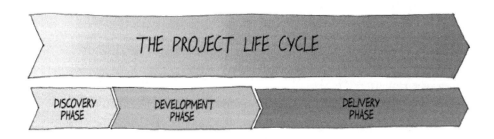

Figure 13.1 CAMMP™ project life cycle, with the three phases.

13.3.3 Stages

As stated, phases are fixed, but projects are not the same, and one-size-fits-all does not work. To accommodate for different projects, CAMMP™ expands the life cycle by one level with the introduction of *Project Stages.* Stages are like phases—they represent a span of time along the project life cycle—but the purpose of each stage is to produce an output, a stage deliverable, which is assessed at a stage gate. Consequently, the scope of each stage is a subset of the phase scope of work. This level of breakdown provides organizations with a greater degree of governance and enables the team to enhance planning and control.

Per Figure 13.2, the current standard version of CAMMP™ now consists of nine stages; there were six in the previous version. The increase in the number of stages, which will also correspond to an increase in the number of gates, is necessary for better clarity. This change came about as a result of different feedback from working with clients and workshops.

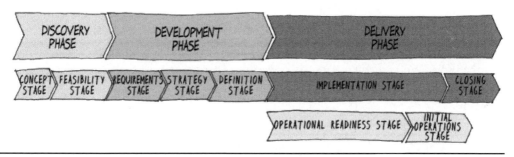

Figure 13.2 CAMMP™ project life cycle, with phases and stages (per the standard model).

Before the reader panics, more stages do *not* mean more work; the same level of effort is required for a given project whether all the work is grouped into three phases only or uses nine stages. The key difference and level of effort is not in the number of phases or stages but in the project type, domain, size, or complexity. With the same number of stages, say nine per the standard model, a small-simple project could be completed in a few weeks or months and with a few people. On the other hand, a large-complex project will require many teams, a large number of individuals, and a duration of a few years.

It is necessary to stress at this time that what is described here is the standard model. The last two parts of this book address customizing and adapting this methodology to different domains, resulting in various models with various numbers of stages and stage gates and a different graphical appearance.

13.3.4 Stage Deliverables and Gates

Even in small projects, there could be many deliverables. *Deliverable* is a term used to reflect a report, study, plan, document, or physical work. A deliverable could be huge and may represent the whole project; a house is a deliverable of a "building a house" project. A deliverable could also be broken down into small deliverables, such as engineering design of the house; or broken down further to civil design or mechanical design for the house. A further breakdown is possible, down to a work package or even an activity.

Section II, Part E will present the stage deliverables in detail. Although there are other deliverables, their detailed explanation is outside the scope of this book. These stage deliverables are the output of every stage.

Once a stage deliverable is complete, it should go through a stage gate. Each stage gate, except the last one, has two objectives: The first target is to confirm whether the stage work is complete per the organizational guidelines and requirements, and the second objective is for the relevant stakeholders to decide if the work is ready to progress to the next stage or not.

In an earlier chapter, another type of gate was introduced: the *process* gates. These are within a stage and are related to the process groups; this type of gate is covered later in the book.

13.4 Summary of Phases and Stages

13.4.1 Discovery Phase

The purpose of the Discovery Phase is for the team to understand and validate the concept for the project and whether or not it is feasible. This is the initial phase and consists of two sub-phases (stages). Just a reminder, the *PMBOK® Guide* and ISO 21500 consider this phase as "pre-project" and not as a part of the project life cycle.

The two stages are:

1. **Concept Stage.** The Concept Stage includes documenting the *idea* for the project, outlining the conceptual business case (justification), determining alignment to strategic objectives, and, if all are acceptable *and* a priority, having management approve the idea and authorize a feasibility study.

   The Concept Stage concludes with a concept paper, idea statement, project brief, or other names for the stage deliverable. This deliverable will be subjected to the first stage gate, and the project either stops here or goes on to the next stage. The CAMMP™ chosen name for the output (the stage deliverable) is *project brief.*

2. **Feasibility Stage.** The main purpose of this stage is to validate the *idea* and to authorize the *project*. Note the intentional shift from *idea* to *project*, which is to emphasize that only at the conclusion of this stage is there a project formally authorized to proceed.

   Validating the idea is accomplished via a comprehensive feasibility study that will consider various factors. The feasibility study is the first formal *and* substantial risk management assessment. Let's keep in mind that projects are usually initiated in response to either a threat (problem) or an opportunity. In other words, projects are risky ventures, and the feasibility study is what helps the organization determine whether to exploit the opportunity, avoid the threat, or mitigate the various risks and pursue the project.

   Once the study is complete, it would be time for another gate. This gate has three objectives, not two.
   - The first answers the question about the study itself and whether it is complete per the organizational guidelines and requirements.
   - The second accepts or rejects the study conclusions.
   - The third is an extension of the second. Accepting the study is one thing, but authorization is another. The organization must consider this project in relation to all other projects that the organization is considering, and authorization will be a function of priority and other factors.

13.4.2 A Message on Authorization

With project authorization, the team can conclude the work of the first phase and move to the next one. However, before moving on, it is vital to highlight the following key point. For proper project governance, authorization at this point indicates the organization's interest and clear intent to go ahead with the project to completion. However, in reality, management should only be approving work for the next stage and phase. In other words, management can still stop the project at the upcoming gates, usually until the final approval, which is normally at the end of the Development Phase.

Today, some organizations decide to approve projects and allocate the budget at this point. This early approval is a choice, but one with significant risks, since the requirements are not entirely clear or developed, and there is no project management strategy or detailed plan.

13.4.3 Development Phase

The Development Phase is necessary for developing the concept of the authorized project to ready it for the Delivery Phase. CAMMP™ splits this phase into three stages:

1. **Requirements Stage.** As the name indicates, this stage identifies and documents the various requirements, including the technical requirements and product characteristics, in addition to the management and project management needs. Furthermore, this stage is where the need for exploring the expectations of the various stakeholders and converting them to clear and explicit requirements. The stage concludes with a project requirements document, which is subject to the Stakeholder Alignment Gate. This gate is critical and, ideally, should help the team freeze the scope, the aim of which is to minimize changes and their considerable impact after this point.

    The author realizes that in certain types of projects, especially software development, fixing requirements may not be easy or desired; Agile approaches may require flexibility, and change would be normal and acceptable. This flexibility would be a special case for special types of projects. However, even in Agile or Scrum, the author still advocates the need for this stage with a focus on the high-level business requirements (refer to Figure 10.2 on page 62).

2. **Strategy Stage.** One can also label this stage as Project Management Strategy or Delivery and Management Strategy, to emphasize the need to develop the project management (and control) plan along with the delivery strategy, especially for projects with some level of complexity. In this context, *delivery* represents the upcoming Delivery Phase. *Refer to the textbox on next page.*

    As before, the output of this stage will be going through another gate. Before the gate, it would be necessary for the project management team to have an updated *feasibility*, with cost and schedule information that can be used to help management decide on continuing with the project or not.

3. **Definition Stage.** The Definition Stage is critical and requires significant effort. The main purpose of this stage is to produce the project detailed plan (PDP), which should

include all of the details for the scope of work, deliverables, applicable quality standards, safety-health-environmental laws (if applicable), detailed schedule and cost estimates, and many other details. It is critical to be as clear as possible with enough detail to allow the effective delivery of the project. Poor planning has severe consequences and is likely to result in the failure of the project to meet some or all of its objectives.

At the conclusion of this stage, the work would have progressed enough to allow relatively accurate estimates, which would enable management to make the final decision to approve or stop the project. Therefore, it is necessary at this time to reconsider the relevant part of the feasibility study to re-confirm the viability of the project.

> **Example of a Delivery Strategy**
>
> In an earlier chapter, we shared a case study about building a petrochemical plant on an island. The study showed that the team considered three approaches for building the facility. That is an example of a delivery strategy and the need to identify it at this point on the project life cycle to account for cost and schedule impact.
>
> In the capital projects industry, this deliverable is often called the *project execution plan*.

13.4.4 Delivery Phase

The Delivery Phase deals with delivering the project per the PDP. The Delivery Phase encompasses four stages (instead of the three in the previous version). Usually, two stages are led by the project management personnel, whereas the operations subject matter experts will lead the other two stages, at least for capital and facilities projects.

The four stages are:

1. **Implementation Stage.** The Implementation Stage requires most of the project effort and resources. It is about performing all of the work packages per the detailed plan. In terms of project management processes, this is a relatively easy stage, but the complexity is due to the significant effort required, depending on the nature of the project.
2. **Operational Readiness Stage**. In parallel to the Implementation Stage, another team should be working on getting ready for post-completion of the physical work. This *operational readiness team* will be developing operational procedures, maintenance procedures (if a physical facility), training for staff on the new systems, and numerous other activities. Obviously, if a project product is not something that will require operations (a facility project), as a training project, then this stage is probably not required and can be eliminated. However, many project types usually require some form of operations or another. A new computer system, a new performance appraisal system, a new building—all are examples of projects that will require operational readiness.
3. **Initial Operations Stage.** In the previous version of CAMMP™, this was part of operational readiness, but we decided to split it out for more clarity. Some facilities projects might require a period that one may call *soft launch* or *commissioning,* which are forms

of initial operations. In other words, this stage is about partial operations or operations at reduced capacity.

Other projects, such as organizational change projects, may require an initial launch in one business unit instead of in the whole organization. This soft launch allows the team to test the product of the project in a controlled environment to learn from it and fix any gaps before full implementation. This work is also called a *pilot*.

4. **Close Stage.** The Close Stage is necessary after all of the work is complete and before the team can claim completion of the project. Successful delivery requires the proper close-out of the project, with all of the necessary activities required at this time. Project closure would include performing the assessment for project success, at least as far as possible at this point; these two dimensions would be product success and project management success. However, at closure, it would not be possible to assess two of the dimensions of project success. Therefore, part of the handover should include a plan as to when the organization can assess success per those two measures along with an evaluation of the realization of benefits. These two dimensions are covered in more detail in Chapter 28.

13.5 The Full Standard Model

After defining the phases and stages, it is time to present the whole picture.

Figure 13.3 presents the standard project life cycle, before any tailoring. The image includes the three phases, the nine stages, and their corresponding stage gates. What this figure does not show are the stage deliverables, but technically at the end of every stage there is output, which is the stage deliverable. These were introduced in this chapter but will be explained in detail in later chapters.

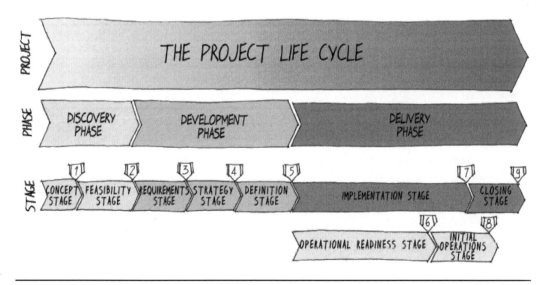

Figure 13.3 CAMMP™ project life cycle, with phases, stages, and stage gates.

It is vital to repeat that *this is the standard model*. Tailoring could include combining stages where possible, eliminating a stage if not required (like operational stages), and possibly subdividing a stage into more than one stage. For example, in many projects, implementation could be broken down into more than one stage, such as design and build, engineering and construction, design, development (coding), and testing, among other possibilities.

However, when it comes to gates, if a team combines stages, it would be necessary to reduce the number of gates, but combining the gates does not eliminate the purpose of any gate.

13.6 The Strategic Aspects in CAMMP™

The Seven Elements of Project Management Maturity™ includes the Strategic and Organizational Aspect as the outer circle. That element includes organizational and project governance and emphasizes the viewpoint that projects must be linked to the organization's strategy and must comply with certain strategic aspects. These strategic aspects, or features, are embedded throughout CAMMP™. The following is a brief summary:

1. The use of the project life-cycle method, which follows a holistic approach from the project owner perspective and covers the span of the project from idea to closure and beyond.
2. The emphasis on including the Discovery Phase, as the first phase of the project life cycle, even though this is somewhat a deviation from ISO and PMI.
3. The insistence on having project management involved during the Discovery Phase, pre-project, and applying the process groups during this phase.
4. The inclusion of stage gates, in particular Stage Gate 1, which focuses on the strategic alignment of the project. What is important to realize is that no project should be approved if it does not align with the organization's strategy.
5. The importance of Stage Gate 2, which is also vital and incorporates the concepts of portfolio management and project prioritization as part of the decision-making process.
6. The inclusion of the Close Stage as an integral component of the Delivery Phase, and the insistence that the project is not delivered until it is properly closed. Including the Close Stage in the Delivery Phase would eliminate a common practice of organization's ignoring project closure and thus not capturing the learning.
7. The view of the four dimensions of project success as essential features. Later, we will introduce a process to define the criteria for success and a process to ensure the measurement of success.
8. Emphasis on benefits management as an integral part of the project and linked to project success.

There are other subtle, direct, or indirect elements contributing to managing projects with a strategic perspective.

13.7 Closing Comments

Just remember these concepts:

- CAMMP™ is a universal methodological approach that integrates the learning from various associations and builds on *proven* practices, not to say *best* practices.
- CAMMP™ application starts with the standard model that organizations must tailor with consideration of many factors, such as project domain, type, size, and complexity.

The chapters in Part E cover the stages in detail, one chapter for each stage.

Section II, Part D
The Second Dimension
(Project Management Processes)

Chapter 14

Links to Global Standards

14.1 Introduction to Part D

Part D is dedicated to discussing the second dimension of the three-dimensional Customizable and Adaptable Methodology for Managing Projects™ (CAMMP™) Model. As this dimension depends heavily on both ISO 21500 and the *PMBOK® Guide*, it would be of value to have this chapter showing the CAMMP™ link to global standards and how they relate to each other.

14.2 CAMMP™ Refresher

CAMMP™ is a project management methodological approach consisting of three dimensions. The first dimension is a project life cycle model covering a given project from idea to closure. The project life cycle consists of many phases and stages; within each phase or stage, a set of processes and process groups applies.

It is worth noting that CAMMP™ is in line with proven leading practices that numerous global large, medium, and even small organizations are using. Furthermore, CAMMP™ incorporates the learning from the various

> **Author Special Message**
>
> We believe that it would be better and clearer if we start with the second dimension instead of the first.
>
> The main reason behind this decision is that many readers are likely familiar with the concepts of process groups, processes, and functions from references such as the *PMBOK® Guide* or ISO 21500, and this approach would ease their transition from these concepts into the more comprehensive and cohesive approach offered by CAMMP™.
>
> It is important to stress that the author still considers the first dimension as the most essential foundation for CAMMP™ or any other proper methodological approach. Furthermore, for small-simple projects, a project life cycle model incorporating key processes would provide a better model than depending on stand-alone processes that are not linked to phases or stages.

global associations—the intention is not to re-invent the wheel but to enhance the practice of project management using the best that exists in the professional community.

Furthermore, it was stated earlier that, although CAMMP™ blends the knowledge from various leading associations, it is not restricted by them. CAMMP™ builds on what these associations offer and fills some of their gaps. Remember, these gaps are not necessarily errors but are due to the mandate of these various associations and their resources. However, some of the gaps could be errors or inconsistencies.

14.3 Professional Associations and Gaps

In order not to repeat, it might be worthwhile for the reader to revisit Chapter 2 and reflect back on Section I of the book. What is listed here are the major gaps that CAMMP™ is trying to fill.

1. One of the main gaps in project management practice today is the belief that a single standard or platform is enough. The challenge is: Will the professional community be receptive to blending or integrating the learning from more than one association? *(Refer to the textbox.)*
2. Organizations need a comprehensive organizational project management (OPM) system, which must include various elements. As discussed in Chapter 9, SUKAD has developed a model for the OPM, The Seven Elements of Project Management Maturity™.
3. A major practice gap is related to the ISO and PMI guides, in which some practitioners think that the process *groups* are project *phases*, which is the subject of Chapter 7.
4. The emphasis of the various global associations on a piece of the puzzle and not the whole picture—for example, IPMA on competence and ISO on processes—is limiting practitioners' abilities to implement proper, comprehensive project management in their organizations.
5. The policy—or mandate—of professional associations to stay generic—"most projects, most of the time" (The Project Management Institute 2017)—can help reach many people but does not provide enough for specialized, industry-specific applications. PMI is trying with the extensions to the *PMBOK® Guide,* but unfortunately, few professionals know these extensions exists.

> **Is the Professional Community Receptive?**
>
> A few years ago, we were awarded a contract to deliver a workshop on program management. The workshop outline followed PMI's *Standard for Program Management,* second edition, and about 50% of the content of the course was in line with it. We did not think that the PMI standard was adequate for proper coverage of program management, so we added the rest of the content from our own work and the work of other program management global leaders, including case studies.
>
> We thought the workshop went well and the feedback in class was excellent. However, later the client representative objected that we added materials from outside PMI. When we explained that, in our professional opinion, the PMI program management standard was not enough to present an excellent learning program, the client was not receptive and took our comment as a criticism of PMI.

6. Finally, the leading gap is that most of the guides of these associations do not include a method or a methodological approach. They mention the need for methods but do not offer one; hence the need for CAMMP™.

14.4 CAMMP™ Link to PMI and ISO

A marketplace reality is that the *PMBOK® Guide* is extremely popular, although it is not the only project management guide available. Although it is valuable, like most standard documents it does not cover everything. It has gaps, inconsistencies, and missing elements, whether by design or oversight. Because of the importance and widespread use of the *PMBOK® Guide*, it is necessary to address how it relates to the CAMMP™ methodological and systematic approach.

ISO 21500 is becoming more noted and used around the world, especially as ISO has started to address project management in recent years.

The processes, process groups, and subjects (knowledge areas) are common features of these two leading guides. The processes, process groups, and functions constitute the second dimension of the three dimensions of CAMMP™. In other words, without them, CAMMP™ is just a project life cycle model limping along. However, as stated earlier, although CAMMP™ aligns with the principles of the various global standards, it is not restricted by any one of them.

In summary, CAMMP™ incorporates the project management processes and functions addressed by these guides and adds to them.

14.5 CAMMP™ Link to IPMA and Other Associations

The third dimension of CAMMP™ includes the following topics:

1. **Competence and Competency Development.** This is the core focus of GAPPS, IPMA, and IPMA members' associations.
2. **Best Practices.** The fundamental concept of best practices is that (1) these practices cannot be covered by guides such as ISO, PMI, or IPMA because of the generic nature of what these associations offer; and (2) best practices have to link to industry-specific best practices. The author has significant experience in the capital project industry, and the third-dimension references the work of the Construction Industry Institute (CII), which promotes best practices in its industry.
3. **Sustainability.** This is the core focus of GPM® (Green Project Management) Global with their P5 and PRiSM™ guides.
4. **Project Success.** This is based on SUKAD Research and Development work and not from any specific professional association.

14.6 How CAMMP™ Differs from Global References

14.6.1 General Statements

How is CAMMP™ different from global references such as the *PMBOK® Guide* or ISO 21500? Are there any deviations in principle concepts? There are some tactical choices listed in the details spread across this book, but following are three categories of difference:

- CAMMP™ does deviate from certain terminology choices, as appropriate and necessary.
- CAMMP™ includes additional concepts that are not found in these guides or are not adequately addressed.
- CAMMP™ introduces a major shift in the planning process group, adds a couple of functions (subjects/knowledge areas), and introduces numerous other processes.

Further to all of the above, the core difference is that CAMMP™ offers a methodological approach that encompasses all of the key concepts from these associations and more.

14.6.2 Terminology Differences

One of the main reasons for using different terminology is the fact that some terms used in these guides might work well for experienced project managers but confuse newcomers to project management. For example:

- The *charter* authorizes a *project* or *phase*.
- The *Initiating process group* starts a *project* or *phase*.
- The *Closing process group* closes the *project* or *phase*.

Inexperienced readers can have difficulty differentiating when the guides are discussing a project or a phase. Because of this confusion, some merge the two concepts and think of them as the same thing, which is one of the main reasons practitioners think the *process groups* are *project phases*. CAMMP™ definitions include:

1. *Function* instead of *knowledge area* or *subject area*. The author chose this term to help the reader avoid the confusion with the 10 knowledge or subject areas of the *PMBOK® Guide* or ISO 21500, respectively. Furthermore, the guides do not include some functions, such as safety and environment.
2. *Project authorization document* instead of *project charter*, to avoid confusion between a project and a phase. In other words, *project* authorization clearly refers to authorizing the *project*; however, when ISO 21500 or the *PMBOK® Guide* states that, "The project charter authorizes the *project or phase*," it creates ambiguity.
3. *Implementing* instead of *executing*, primarily to differentiate between *implementation as a stage* and *executing as a process group*.
4. *Project Detailed Plan* instead of *Project Management Plan*. This is not just a terminology difference, because these two deliverables are not the same, as outlined earlier.
5. *Phase* and *stage* are differentiated, whereas most of the global guides use these terms interchangeably.

14.6.3 Additional or Emphasized Concepts

CAMMP™ includes additional concepts that are not found in these guides or are not adequately addressed. These include:

1. **Stage Gates.** The CAMMP™ approach emphasizes a strong stage-gate process as an essential control feature; this lacks emphasis in the guides, as evidenced by the absence

of gates in any figure in these references. For example, earlier editions of the *PMBOK® Guide* have ignored this concept, but finally, the sixth edition includes a brief mention of gates, and even some of the guide's figures show gates. Furthermore, in the guides, the stage gates are implied in the initiating and closing process groups. In other words, initiating or closing a stage requires a decision, and the stage-gate reviews are where management makes the major decisions.

2. **Discovery Phase.** The guides consider that the early, pre-charter project effort is usually outside the project life cycle and define the project life cycle from charter to closure. CAMMP™ considers the project life cycle starting with the idea and project brief.

3. **Phases.** The guides do not name the phases, since they vary from one domain to another. CAMMP™ names were selected to reflect the scope of each phase or stage, with the understanding that alternative names are acceptable.

4. **Project Charter.** The guides use the term *project charter* to refer to the "document that authorizes a project or phase." To avoid confusion, in CAMMP™ the document that authorizes the *project* has a name that is unambiguous: *project* authorization document (PAD). Furthermore, through the project life cycle, phase and stage authorizations take place at stage gates, where executive management grants approval and is in effect authorizing the next stage. These are called *stage* authorization documents.

5. **Project Management Plan.** One of the main differences between CAMMP™ and the guides is the use of the term *project management plan*. What the guides call a *project management plan* and *project documents*, CAMMP™ splits into two documents (or set of documents) produced in two different stages: (a) the *project management plan* (PMP) developed during the Strategy Stage; and (b) the *project detailed plan* (PDP) developed during the Definition Stage.

> ### The Stage-Gate Process
>
> "A prime area executives should concentrate on to improve capital projects is the stage-gate process. . . .
>
> "IPA has definitive data showing that projects that follow the stage-gate process have much less deviation in delivering the opportunity value promised to shareholders.
>
> There is no getting around the fact that executives benefit from following a stage-gate process for delivering projects, but the process itself is not self-sustaining. It is up to executives to mandate that the project sponsor and team follow the process."
>
> Source: http://www.ipaglobal.com/the-executive-role-in-making-capital-projects-pay-off

6. **Project Success and Benefits Realization.** These topics are either missing or barely mentioned in the guides until the sixth edition of the *PMBOK® Guide*, but for CAMMP™, project success is a critical component of project management, and benefits realization is also important to the CAMMP™ Model.

14.6.4 Shift in Planning and Additional Processes

CAMMP™ introduces a major thinking shift in the planning process group and introduces numerous other processes:

1. CAMMP™ offers two planning process groups—the management planning group and the detailed planning group—which is a deviation from the guides and even the PDCA Cycle.
2. Over the years, the number of processes in the *PMBOK® Guide* has been between 37 and 47, and ISO 21500 has 39 processes. The number of processes is not necessarily a function of the effort, since one can consolidate these processes into less than 20 processes or split them further into 100 processes and subprocesses. The effort is linked to what has to happen to produce an output. Smaller processes with a narrow focus can be developed more quickly than long processes with bigger outputs. For example, per the *PMBOK® Guide,* planning is a process group consisting of 24 processes. However, one can also view planning as a process (a single high-level process) comprising subprocesses, one for each knowledge area, then sub-subprocesses within a knowledge area. For example, developing the schedule is currently five subprocesses; can they not be combined into one process? Alternatively, they can be broken down into seven or eight subprocesses.

With this introduction in mind, CAMMP™ is taking an approach of smaller, focused, and clear processes, and they are explicit, not implicit. Chapter 17 introduces the modified view on processes and offers the additional processes (or subprocesses) that are part of CAMMP™.

14.7 Closing Comments

CAMMP™ is a methodology that must be part of the organizational project management (OPM) system. The first dimension is a project life cycle model, which is not enough to manage projects effectively unless they are small–simple projects. Therefore, it is necessary to include project management processes and subprocesses as the second dimension. The third dimension utilizes concepts from other standards. The blending of these various concepts is what provides the project management professional community with a robust, systematic, and comprehensive approach.

Therefore, CAMMP™ is not intended to criticize, contradict, or replace the substantial knowledge base that exists today, whether that knowledge is from professional associations, organizations, universities, or thought leaders. The intent and aspiration of CAMMP™ are to be a significant value addition to the emerging field of *applied project management.* It intends be the methodological frame to implement the learning and glue the vital pieces of the process, method, competence, sustainability, and other topics to complete the project management puzzle and enhance organizational performance.

Chapter 15

Management and Detailed Planning

15.1 Introduction

Planning is a challenging topic for many reasons. It is time to address it in depth, because what CAMMP™ incorporates is a major change to the current conventional wisdom of one planning process group.

Here is the current reality, inclusive of misunderstandings and corrections.

- Some practitioners think planning is a phase, yet planning is about repeated processes that occur in every phase.
- Other practitioners believe there is only one plan per project—the project management plan, which is the output of the planning process group.
- However, there should be at least one management plan for every phase or stage, as the planning steps do repeat. (Refer to Chapter 7.)

> **ISO and the PMBOK® Guide**
>
> Earlier in this book we stated that, ideally, one should compare ISO 21500 to the *PMBOK® Guide* Part 2, which is the ANSI-approved standard. It would not be proper to compare ISO to the whole *PMBOK® Guide*.
>
> A key difference is that the *PMBOK® Guide* includes input, tools and techniques, and outputs for all processes, whereas ISO 21500 does not.

Consequently, if practitioners accept the ISO and PMI guides, then by default they are accepting the concept that *there is at least one plan per phase*. This plan is the output of the planning process group.

Here one must ask, is there one plan *within* a phase, or could there be more?

15.2 One or More Plans?

The *PMBOK® Guide* has a process called the *Develop Project Management Plan,* which is for the project or phase. ISO 21500 has a similar process called *Develop Project Plans.* It is interesting to see the difference in the chosen terms—one uses *plan* while the other uses the plural form *plans,* and one focuses on *project* management, the other on the *project.*

However, if one studies these guides' planning processes and understands them well, one will notice that there are *two types* of planning processes: two sub-groups. These are *management planning* processes and *detailed planning* processes. Consequently, each set of these processes produces a management plan and a detailed plan—this is the CAMMP™ premise. On the other hand, the ISO 21500 planning processes focus on the detailed processes. This is reflected indirectly but clearly in the number of planning processes: the *PMBOK® Guide* has 24, whereas ISO 21500 has only 16. ISO does not have a *Management Planning* process for each subject area.

15.3 Splitting Planning

We had mentioned in earlier parts of this book that CAMMP™'s focus is on project management and delivery and is not limited to project management. Consequently, for every project, we must distinguish between *product* scope and *project* scope; product scope would be about the final result or output that the project delivers. Furthermore, the team must also distinguish between *project* work and *phase* or *stage* work.

In other words, every project has an objective, which is to deliver an output or a product with a clear product scope. The project team will follow a project life cycle approach that will lead to the desired outcome by going through the steps and completing the project work. The project work, along the project life cycle, consists of phases and stages, each with its own purpose and scope of work. Therefore, it is vital to think about product, project, and phase work and understand the differences between them.

Based on the above, planning is a vital process to help the team travel a project journey. Therefore, it is important to segregate the management effort from the technical and functional planning, and for planning, the CAMMP™ approach is per the following:

- There will be two plans for the whole project—a project management plan (discussed in Chapter 21) and a project detailed plan (explained in Chapter 22).
- Each project phase or stage, except for simple projects, will also have two plans—a stage management plan and a stage detailed plan; both are discussed in the next chapter.
- Usually, the project management personnel will develop the project and phase-stage management plans, whereas other project personnel (technical and functional resources) will complete the work required for the project and phase-stage detailed plans.

15.4 Closing Comments

Once again, the key message in this chapter is to think about the big picture *and* the details, the product versus project and project versus phase-stage. What we need is a mindset that expands the focus from *managing* projects to *delivering* products successfully, which would require a changing paradigm.

Chapter 16

Processes and Functions

16.1 Introduction

CAMMP™ uses the term *functions* to refer to areas of specialties in project management. ISO 21500 uses the term *subjects,* and *PMBOK® Guide* refers to them as *knowledge areas.* Per the current editions of these guides, 10 generic functions would apply to "most projects, most of the time" (The Project Management Institute 2017).

Certain functions might be limited to one domain or another, and it would make sense not to include them in a generic guide. ISO does not have any guides for particular domains, but PMI does have supplements to the *PMBOK® Guide* that would be industry specific. PMI labels these supplements as extensions to the *PMBOK® Guide.*

Another change in the terminology is the term *process groups.* The PDCA cycle had four components, ISO 21500 and the *PMBOK® Guide* each has five sets of processes called process groups. Each group consists of two or more independent processes, except the closing process group. For these guides, these process groups apply at the project or phase level.

CAMMP™ takes another perspective, which is in two parts:

1. For the project level, the project team follows a project life cycle with a set of project phases and stages, as outlined in Chapter 13.
2. At the lowest level of the project life cycle—the stage—the team will follow a set of six processes.

Section 16.2 expands and explains the six project management processes, and Section 16.3 addresses the project management functions.

94 ■ Project Management beyond Waterfall and Agile

16.2 Project Management Processes

16.2.1 The CAMMP™ Processes

In line with the concepts presented in the previous chapter, the CAMMP™ approach is to split planning into two processes—management and detailed. Consequently, there are six main processes that will apply at the stage level, and each process has an output: a deliverable. Figure 16.1 illustrates this concept, and following is a brief explanation of the key points:

- The solid lines show the normal flow of activities within the stage.
- The normal flow starts with Authorize. Then the team will move through the steps in the sequence shown. Notice that Plan Management must come before Plan Details.
- Also, notice the two solid arrows linking Control with Implement; this indicates normal and ongoing flow, going both ways. This graphical designation represents that, throughout the work in progress, the team would be implementing and controlling, with information flowing in both directions.
- The two-sided dashed lines also reflect a two-sided flow, although not as significant as in the case of Implement and Control. This concept represents that the dual flow is ongoing, but it is mostly qualitative rather than quantitative.
- All of these two-sided dashed lines connect with Control. It is represented this way because Control starts once the stage is authorized and applies in all other processes.
- The small circles with PG1, PG2, PG3, and PG4 are process gates.
- Although the work should be somewhat sequential as shown here, it is possible to have some overlaps to fast track, but this will increase the threats (risks).

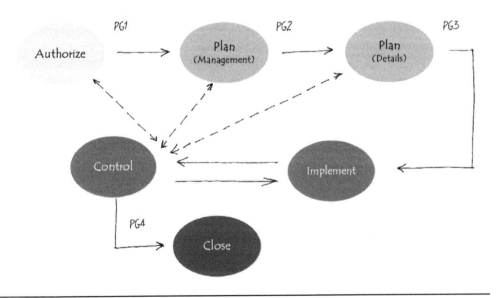

Figure 16.1 The six processes per the CAMMP™ Model.

16.2.2 Process Gates

Earlier in this book, we stated that there are two types of gates: *stage gates* and *process gates*. There are nine stage gates along the project life cycle, with one stage gate at the end of every stage (covered in Part E). Within each stage, the processes repeat, and for some processes, there will be a primary deliverable that must pass through a process gate.

For example, the stage authorization document (SAD), an output of the Authorize process, can be prepared by a project manager, but it has to be approved by the sponsor at Process Gate 1 (PG1). The approval of the stage management plan (SMP), the output of Plan Management, is at PG2, and the stage detailed plan (SDP) approval is at PG3. Finally, before closing, there is PG4 to confirm that all of the stage work is complete before closing the stage. These four process gates will repeat at every stage, just like the corresponding processes.

16.2.3 Typical Process Within a Stage

Figure 16.2 (on next page) presents an illustrative, high-level process flowchart for a given stage. Unless there are special conditions, most stages will follow this example. The only difference from one stage to another is the *work,* the scope of the stage, which changes from stage to stage since the work is product oriented.

16.3 Project Management Functions

16.3.1 CAMMP™ Perspective

The CAMMP™ perspective is per the following statement: because organizations and project managers can select the appropriate processes to their projects, then why not consider functions not covered by ISO or PMI?

Therefore, the CAMMP™ view is that safety and environmental aspects apply to many, if not most, projects. It is recognized that on capital projects, safety and environmental aspects can be critical and more important than in organizational change, software, or technology projects.

However, the following might be two cases to justify including them.

- A few years back, computer batteries for a certain manufacturer were deemed dangerous and a potential hazard due to overheating and catching fire.
- In 2016, a leading mobile manufacturer recalled a recently released popular cell phone and lost millions, if not billions, of dollars due to exploding batteries. This situation was so drastic that most airlines banned these devices on their flights.

Along the same lines, if one separates the concept of environmental aspects from sustainability, would not environmental matters apply to many projects outside the capital projects industry? As an example, projects can be notorious for producing an abundance of paper, which is not environment friendly.

The project management process within a typical stage

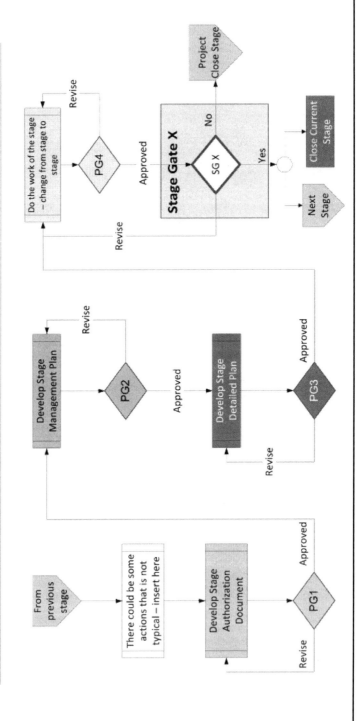

Figure 16.2 A typical process flowchart for a given stage, the main processes.

Finally, health. Most projects consider health aspects, although health matters might be more necessary in one domain than another. Capital projects—projects in hospitality, food and drink, healthcare—all exist in domains in which health considerations are vital.

Therefore, CAMMP™ considers safety, environment, and health as functions required on projects. On a given project, if these topics are not applicable, then the team can categorize them as N/A (not applicable) and skip them. These themes can be grouped into one function, HSE (health, safety, and environment), or split into three functions. Let's keep them together for now.

In addition to HSE, resource management is another topic that should be included and is necessary for managing projects. ISO has this subject, and the sixth edition of the *PMBOK® Guide* replaced the human resource topic with resource management. For certain projects, resource management (other than human resources) can be a major topic. Therefore, similar to HSE, resource management is added in the CAMMP™ Model as another function *independent* of human resource. In the sixth edition of the *PMBOK® Guide,* teams and other resources (physical) are combined in one knowledge area; CAMMP separates them and assigns a dedicated function for each topic.

Another consideration was to combine HSE with sustainability, but they are kept separate. The main reason for the separation is that HSE is mostly about compliance with laws and regulations, so these are often mandatory functions, whereas sustainability (1) is a much wider topic; and (2) is about organizational responsibility and going beyond minimal compliance—although some aspects might be becoming mandatory. Therefore, HSE and sustainability are separate topics; organizations can choose to combine them. Furthermore, in this book, sustainability is considered an advanced topic and is maintained within the third dimension, rather than the second, which is specific to processes of the various basic functions.

16.4 Closing Comments

The author realizes that the content of this chapter may challenge the conventional wisdom for some readers. One can understand that those professionals working on small–simple projects might think this is too much. For the small–simple projects, one can consider a method such as CAMMP™ in its basic form, using only the first dimension and ignoring the repetition of the process groups for each stage.

Medium- and moderate-complexity projects might require the repetition of the process groups, but only at the phase level, not the stage level (refer to Chapter 41). However, for larger and more complex projects, the above approach is needed—and it is vital. It is also important to note that working capital projects already use such an approach.

Chapter 17

Processes and Deliverables Within a Stage

17.1 Overview

Chapter 14 included a mention that, concerning the number of processes, one could have very few large processes with a wide reach, or many small processes with a narrow focus. The The Customizable and Adaptable Methodology for Managing Projects™ (CAMMP™) approach consists of six processes for each stage, and each process includes more than one topic or step.

Four of the six processes produce a deliverable for the stage. Table 17.1 presents the CAMMP™ processes and corresponding deliverables and gates.

Table 17.1 Stage Processes, Deliverables, and Gates

CAMMP™ Process	Deliverable	Process Gate
Authorize	Stage authorization document	Process Gate 1
Plan (Management)	Stage management plan	Process Gate 2
Plan (Details)	Stage detailed plan	Process Gate 3
Implement	No specific document, but the deliverables are the completed stage work	See Control; the Control process will confirm completion of the work
Control	No specific document, but there could be many items and reports	Process Gate 4
Close	Stage close-out report	No gate[a]

[a] The closure of all stages will be required for the Close Stage, which has the final stage gate, SG9.

To produce each of the above deliverables for a given stage, the project team will have to complete various steps, which would be documented via the various deliverables and verified at the process gates. In the remaining sections of this chapter, we will address each process. See Table 17.2 on page 102 for a list of steps in each process.

17.2 Authorize Process

The first process is the Authorize process, which leads to the stage authorization document (SAD). In support of the stage authorization document effort, there are four key steps and actions to complete, and they are discussed next.

17.2.1 Develop Stage Authorization Document

The SAD includes all of the information necessary to authorize a stage. It is a must for the first stage. However, for the subsequent stages, some organizations may choose to combine it with the stage gate documentation for the gate that authorized the stage.

The level of effort for a given project is a function of the size, complexity, and importance of the project, not a function of how many processes or subprocesses it contains!

17.2.2 Establish Stage Success Criteria

Chapter 28 is dedicated to project success, which is per the SUKAD Four Dimensions of Project Success Model™; these are technical success, project management success, project delivery success, and objective success. However, at the stage level, the team must also define the stage acceptance criteria (part of the authorization) and stage success criteria. Therefore, defining the success criteria for the stage is in the Authorize process at every stage. Establishing the success criteria for the project can be within the project authorization document (PAD).

One key difference between stage success and project success is that at the project level, CAMMP™ depends on the four dimensions; but for the stage level, there are only two dimensions—*technical success* and *stage (project) management success*.

One more comment: It would be good to show a link to the organizational project management system (OPMS) at this point. For organizations performing similar projects, some of the project success dimensions, metrics, and key performance indicators could be standardized and included in the OPMS. Therefore, CPM (copy-paste-modify) would be enough. It is as simple as that.

17.2.3 Mobilize Project Manager

For some projects, selecting a project manager could be an automatic selection and would not require any effort; in that case, there is no need for a dedicated process. In that situation, the process would not be applicable. However, in other circumstances, selecting and mobilizing a project manager can be a major effort. Therefore, choosing the project manager deserves a dedicated focus, and CAMMP™ includes this step.

What could happen to this step when the project moves into the second stage?

It is simple: If the project manager will continue on the project, then there is no need to repeat this step. However, if the project manager is changing, then it is time to apply this action again.

17.2.4 Identify Stakeholders

As the sponsor authorizes the stage and appoints a project manager, it is essential to identify all of the relevant stakeholders for the given stage, especially in the first stage. For subsequent stages, it is still important to revisit this step, because some stakeholders might no longer be relevant, whereas others could be added and will have a role or impact.

17.3 Plan Management Process

The Plan Management process per the CAMMP™ Model will include seven steps/topics to produce a stage management plan (SMP).

First, the project manager starts the effort by outlining the stage management plan and all of the topics that the team must address. Next, the project manager assesses the effort required for planning and managing the stage to identify the roles and number of project management resources needed and mobilize them.

The project management team, working with the project manager, will proceed to plan how to manage the stage work, considering every function that is applicable. Some of the items and actions to consider are planning for how to manage change, quality, risk, cost, scope, resources, HSE, etc. During this planning effort, the project management team will define the resources required for the Plan Details process.

17.4 Plan Details Process

The Plan Details process is necessary to produce the stage detailed plan (SDP). This is a comprehensive plan that includes a great number of details on all of the required functions for the stage. The SDP includes the detailed work scope; estimates of cost and time, resources, and staffing plans; quality and HSE compliance details; risk assessment and treatment; purchasing and contracting scope and requirements; etc.

To develop the details for all of these functions, the team is likely to use 20 independent process steps.

17.5 Implement Process

The Implement process is straightforward and is mostly about performing the stage work and completing the stage work packages per the SDP and in accordance with the various specifications and compliance requirements identifed earlier.

Table 17.2 CAMMP™ Processes Step by Step

The 4 Steps of Authorize Process	The 9 Steps of the Implement Process
Develop Stage Authorization Document	Complete Work Packages
Establish Stage Success Criteria	Mobilize Implementation Team
Mobilize Project Manager	Manage Team
Identify Stakeholders	Manage Stakeholders Engagement
	Manage Communications
The 7 Steps of Plan Management Process	Acquire Resources
Develop Stage Management Plan	Select & Mobilize Service Providers
Develop Change Management Plan	Manage Contracts
Develop Control Plan	Lead & Manage Stage Work
Develop Configuration Management Plan	
Develop Stage Initial Schedule	**The 16 Steps of Control Process**
Develop Sage Initial Cost Estimate	Control Scope
Mobilize Management Team	Control Schedule
	Control Costs
The 20 Steps of Plan Details Process	Control Risks
Explore for Expectations	Carryout Quality Assurance
Create Stage PBS & WBS	Control Quality
Develop Stage Scope of Work	Carryout HSE Assurance
Develop Stage Schedule	Control HSE
Optimize Stage Schedule	Control Team
Develop Stage Cost Estimate	Control Stakeholders Engagement
Determine Stage Budget	Control Communications
Identify & Clarify Risks	Control Resources
Assess Risks	Control Contracts
Plan Responses to Risks	Control Stage Work
Define Quality Standards	Control Changes
Define Process Improvements	Assess Performance
Define HSE Compliance Requirements	
Mobilize Detailed Planning Team	**The 8 Steps of Close Process**
Determine Implementation Roles and Staffing Requirements	Validate Product
	Demobilize Team
Develop Stakeholders Engagement Plan	Close-Out Report
Determine Communications Requirements	Close Resources
Determine Resources Requirements	Close Contracts
Develop Procurement Documents	Document Lessons Learned
Outline Stage Detailed Plan	Reconcile Stage Work
	Assess Stage Success

The first step is to mobilize the implementation team with the right skill sets for the given stage work. It is likely that the implementation team will change from stage to stage as a function of the type of work.

To accomplish the output of the stage, the project manager will lead the management effort and function as an essential link between the team on one side and the sponsor and management on the other. While implementing, there could be nine steps and actions.

17.6 Control Process

The Control process consists of numerous steps and actions. In general, every function will have a plan to manage it,[1] a detailed plan to cover the essential details,[2] and control steps. Therefore, Control includes scope control, cost and time control, resources and team control, contract administration, in addition to all of the other functions.

17.7 Close Process

The CAMMP™ project life cycle includes a Close Stage to close the whole project. The Close Stage will depend on the Close process that would be required in every stage. Closing a stage includes the following steps: validate that the stage work is complete, measure stage success, close any open contracts or purchase requisitions, complete the close-out report including lessons learned and performance reconciliation, and, finally, close the team—adjourn.

17.8 Closing Comments

This has been another extensive chapter. It is crucial to remind the reader that effort expended on a project is a function of project size, importance, and complexity, not a function of the number of processes. In the end, a process is any work required to produce an output. A narrowly focused process can be accomplished quickly, whereas a large, wide process will take much longer.

CAMMP™ limits the number of processes to six per stage, and they do repeat from one stage to another. However, each of those processes consists of many steps (or subprocesses), and there could be up to 64 process steps within a stage. This approach is necessary for clarity and to ensure great project performance.

This chapter closes Part D and ends the discussion on the second dimension of The Customizable and Adaptable Methodology for Managing Projects™ Model.

---

[1] Part of the stage management plan.
[2] Part of the stage detailed plan.

Section II, Part E
The First Dimension
(Project Life Cycle – Details)

Chapter 18

Discovery Phase – Concept Stage

18.1 Phase Overview

The Discovery Phase is for the team to understand and validate the idea for the project and whether or not it is feasible. It is the initial phase and consists of two sub-phases (stages): the Concept Stage and the Feasibility Stage. Figure 18.1 is an overview of the process sequence for the Discovery Phase, incorporating the main processes. The first part, in the top left corner, is the Concept Stage, which is quite limited in scope and is only deliverable is the project brief (covered in this chapter). The rest of the image is the Feasibility Stage, which is covered in the next chapter.

What is not shown in Figure 18.1 is the phase and stage control process, because all controlling steps are happening in parallel to all of the shown steps from start to finish.

18.2 Stage Overview

The Concept Stage includes:

- Documenting the *idea* for the project, project description, or statement of work
- Outlining the conceptual business case (justification)
- Determining alignment to the organizational strategic objectives
- If all are acceptable *and* a priority, management's approving the idea and authorizing a feasibility study

Some of the titles for the key stage deliverable are *venture statement, project idea statement, project brief,* or any other suitable name. CAMMP™ adopts the term *project brief* (PB). This stage

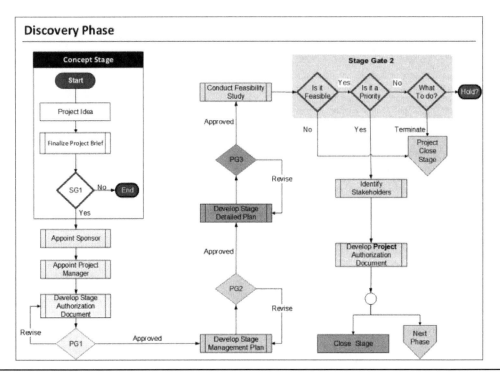

Figure 18.1 Discovery Phase, process overview.

deliverable will be subjected to the first stage gate, and the project either stops there or goes on to the next stage.

18.3 Sequence of Events

The sequence of events for this stage is simple and would be per the following:

1. Any project starts with an idea, a driver, a trigger, or whatever else one might call it.
2. The idea will be documented and submitted to management.
3. Management will review and decline, approve, or require more analysis or rework.
4. If declined, the work must stop.
5. If approved, management can either hold for the future or authorize a feasibility study.
6. If it is a go, management should appoint a project sponsor and a project manager to lead the work in the upcoming stage—the feasibility study.

18.4 Special Note About Processes

Before continuing, we should emphasize and remind the reader that *project management effort is a function of the project size, importance, and complexity*. For small–simple projects, the effort required for initiating and planning the stage might be limited or almost negligible. It may even be verbal or through emails with no formal documentation. On the other hand, if the projects are moderate in size, large, or complex, the effort can be substantial and require a high level of resources, effort, and time. Therefore, the team must consider applying the processes for each phase or stage.

For this first stage, it would be safe to consider skipping the application of the CAMMP™ six main processes, except for large–complex projects. For these projects, the Concept Stage could require significant effort, and it would be similar to a project in itself, justifying the application of any applicable project management processes.

18.5 The Project Brief

18.5.1 Overview

The idea for a project could come from anyone within the organization. In theory, there are two sources for ideas.

1. The first source is *top down,* meaning all projects must be initiated from the strategic planning process, because the strategic plans must have strategic initiatives, which executives would convert to portfolios or strategic programs and, in turn, projects. This approach is the ideal scenario for organizations that truly follow a strategy formulation and implementation process. In this scenario, every project does align to the organization's strategic direction, which is a requirement for the project brief.
2. The second source is *bottom up.* Ideas could come from an employee, a strategic business unit, a department, or executive management. It could be a new concept or a modification of something existing. Organizations should have a process for idea generation, review, and disposition, which could be part of the business planning process or strategic planning. Some organizations could even have an innovation management process in which ideas could come from anywhere, including from customers.

Regardless of what idea management system exists, the idea for a new project must go through an idea management system or project approval process. In small organizations, the owner of the idea might be able to present the idea directly to senior or executive management. In larger organizations, there might be layers of reviews. However, the preference would be that the owner of the idea should be able to present their initiative to senior management or the decision makers, wherever possible (SUKAD 2016).

The idea presentation must be clear, define the key concept, and identify the strategic value to the organization. The idea definition and presentation must be brief. Management could require additional work before deciding on an idea, which is normal.

18.5.2 Potential Pitfall

The organization must be careful not to fall into the trap of either spending too much time on idea definition and clarification, or confusing the work required for the project brief versus what the organization should do in a feasibility study or even at later stages.

In well-established and mature organizations, executives understand that at the time of the idea introduction, no one, including the idea originator, has all the answers, and some ambiguity about the details is not only accepted but also expected. What this means is that the idea generator should not plan the project *before* presenting the idea. This would not be a sign of a healthy development of the project brief; rather, it is a clear indicator of system failure, wherein team members, the idea generator, and possibly others are using company resources to work on projects that management has not authorized.

The key here is to perform enough work to understand the idea so that management can decide on a path forward—no more, no less. Of course, the level of detail is a function of the degree of empowerment within the organization and the organizational culture.

18.5.3 Content

At times, clients and colleagues ask us for a template or form for idea documentation. Our answer: it is just a short paragraph, a few sentences or slides. If the idea is not clear in a few minutes, then the idea generator has failed in presenting the idea. The project brief should include the idea and the driver behind it—the justification. In addition to the *what and why*, it is necessary to include alignment to the strategic direction!

There are only three points to address:

1. What is the project all about? A short description of the product; the output of the project.
2. Why should the organization pursue, or continue to pursue, this project? The justifications and the expected outcome.
3. How does the project align with the organization's strategic objectives? In organizations with a strategic plan, each project must align with one of its goals.

18.6 Stage Gate 1 (SG1)

> *There is nothing as wasteful as doing with great efficiency that which does not have to be done at all.*[1]
>
> — Peter F. Drucker

It is time for bringing strategic project management into the discussion.

This first stage gate, SG1, is about approving the project brief, but it is primarily about strategic alignment. *Every project must be in line with the organization's strategic direction and objectives.* If organizations do not consider this strategic view as a priority, then project management would not be highly effective in contributing to long-term organizational success (see quote above).

There are side benefits to the first stage gate's being about the project alignment to the organization's strategic objectives. Let us consider the following:

- If strategic alignment is an ingrained principle in the culture of the organization, and
- Professionals within the organization have a *line of sight* on how their roles and work contribute to organizational success, then
- The idea originators would suggest ideas that will help the long-term success of the organization, instead of wasting time on ideas that would not align with the core business.

---

[1] https://www.goodreads.com/quotes/29838-there-is-nothing-quite-so-useless-as-doing-with-great

In other words, projects are value-improving efforts in organizations, and management should consider them carefully.

In closing, who approves the project brief?

This is a function of organizational size and culture. In some organizations, empowered line managers or senior executives might be able to approve up to certain pre-authorized limits or other predefined criteria. For major projects, this could be executive management, the board of directors, a government ministry, or any other high-level decision maker.

18.7 Stage Summary and Next Steps

With approval at SG1, the organization will appoint a project sponsor to assume leadership for the project. The project sponsor is senior management's representative and would be the person responsible for the project overall, but not the day-to-day project management (Englund 2006).

The author advocates that the organization should have project management involved at this point. The sponsor, or management, should appoint a project manager to lead the upcoming work. The project manager's focus is on managing the day-to-day work, whereas the sponsor's is to maintain the business and management perspective.

Chapter 19

Discovery Phase – Feasibility Stage

19.1 Stage Overview

The main purpose of this stage is to validate the *idea* and to authorize the *project*. Note the intentional shift from *idea* to *project*, which is to emphasize that only at the conclusion of this stage is there a project authorized to proceed.

Validating the idea is accomplished via a comprehensive feasibility study that will consider various factors. There are two key stage deliverables: the feasibility study (FS) and the project authorization document (PAD). There is a stage gate at the end of the study, which will lead to the PAD if the project is approved. Figure 18.1 (on page 108) presented the process flow.

19.2 Sequence of Events

The sequence of events for this stage is simple and would be per the following. Please note that the same general sequence will apply for the upcoming stages.

> **Project Management Team**
>
> The project management team (PMT) consists of the personnel that will support the project manager in managing the stage and project work. The team members may include planners; cost, schedule, and procurement specialists; and other personnel as required.
>
> The PMT could be limited to the project manager in small–simple projects or include numerous resources for larger projects.
>
> For large–complex projects, the PMT will start with a few people in the first stage of the project and grow as the project proceeds into the other stages, peaking during the Implementation Stage.

1. The project sponsor will authorize the project manager to start working. This authorization could be formalized per Stage Gate 1 (SG1) documentation or through a separate stage authorization document. The stage authorization is equivalent to a stage charter.
2. The project manager will identify and mobilize the necessary project management personnel to help in planning the stage via developing a stage management plan. *(Refer to "Project Management Team" textbox on previous page.)*
3. Once the stage management plan is complete, there is likely a need for a staged detailed plan. The PMT and other lead specialists with expertise in feasibility studies will develop the stage detailed plan. The project manager, project management personnel, and leads will be the core project team (PT).
4. Once the team finalizes the stage detailed plan, the PMT will identify and mobilize the feasibility team—the technical and functional personnel who will be conducting research and developing the feasibility study. This action will lead to acquiring the remainder of the team members to complete the project team for the stage. *Stage team* would be a more appropriate term, but we will use the term *project team* to represent the stage teams and overall project team.
5. The PT will perform the feasibility study and prepare the final report with their recommendations. The work will be carried out by the technical and functional personnel but managed by the project manager and supported by the PMT members.
6. The PT will submit the report to management for review and decision making at SG2.
7. If executives approve the project at Stage Gate 2, the project sponsor will issue the project authorization document (PAD), which is the authorization for the whole project and is equivalent to the project charter. Obviously, if no approval is given, the project would be stopped and closed.

> **Technical & Functional Personnel**
>
> The technical and functional personnel are likely to change from one stage to another. These are the personnel that will do the technical or product-related processes and work.
>
> For example, for the feasibility study, the project may need experts in market research, sustainability, financial modeling, economy, etc. The number and the disciplines of these resources would be a function of the project domain and type.

19.3 The Feasibility Study

19.3.1 Introduction

Let's start with clarifications of a few terms.

1. *Project brief* (PB) is the CAMMP™ term that includes the statement of work (what will the project produce), the justification for the project (why is the project necessary), and the strategic alignment (how does it add value to the organization).
2. *Business case* is a common term but lacks clarity. Some use it to mean the project brief (as explained in the first point); some view it as the justification (the reason for the project

only); and others expand the business case to mean the full feasibility. The latter definition relates to the view that the business case is only valid after the viability of the project is determined. Per CAMMP™, the definition of the business case is related to justifications and strategic alignment. Therefore, the PB incorporates the business case, and there is an independent feasibility study.

3. Even though many projects could have a strong business case (need), at times that is not enough to approve a project, because the organization might not be able to deliver it successfully because of a variety of factors. These factors could be resources, expertise, or something else. Therefore, the CAMMP™ viewpoint is that organizations should pursue a project if it is justified (there is a business case) and the organization's management believes that the project can be delivered successfully and realize the expected benefits. That is the purpose and definition of the feasibility study and why it is a critical stage and deliverable. Organizations cannot afford to skip performing a proper feasibility study —it would be a waste of effort and resources if the organization only discovers that a project is not feasible during project development or delivery!

> **New Terms: Project and Stage Levels**
>
> The reader might be surprised at so many new terms, such as *stage charter, stage authorization document, stage management plan, stage detailed plan, project and stage team,* and others.
>
> Remember, the processes repeat, and there will be specific deliverables related to the main processes that apply within a stage, such as the ones mentioned here. There will also be deliverables for the overall project.
>
> In other words, this book offers two types of deliverables and gates: (1) stage deliverables and stage gates that would be discussed in this part of the book, because they are related to the project life cycle; (2) deliverables and gates within a stage, which were discussed in the previous parts.
>
> For example, there is a *stage* authorization document and a *project* authorization document. Similarly, there is a *stage* management plan (within a stage) and a *project* management plan (along the project life cycle). The same applies to other main deliverables.

19.3.2 Feasibility Study and Risk Management

Let's keep in mind that projects are usually initiated in response to either a threat (problem) or an opportunity. In other words, a project is a risk event, and the feasibility study is what helps the organization to determine whether to exploit the opportunity, avoid the threat, or mitigate the various risks and pursue the project.

19.3.3 The Content

What are the main factors to consider in a feasibility study? Figure 19.1 is a list of the different factors and content of the feasibility study. However, please note:

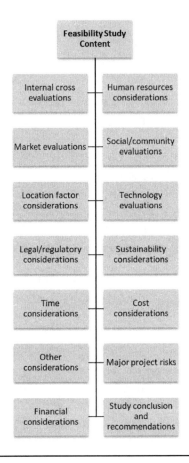

Figure 19.1 Content of the feasibility study.

- Not all of these factors would apply to all projects.
- It would not be possible to list all factors for all projects. Therefore, this list might miss some factors—hence the inclusion of the "other considerations" element.
- The real factors to consider would depend on the type of project and its size, complexity, and domain.
- As the team tries to perform the various activities for each of the factors, they should do so with a risk-management–driven mindset. Meaning, they should think of the things that would be threats to the project's objectives or opportunities that can have a positive impact.

The feasibility factors to consider are discussed below. The team can perform some of them in parallel; others require input from the other factors and would be wholly or partially sequential.

Internal Cross-Evaluation

The internal cross evaluation is necessary because in medium to large organizations, there might be a need to search internally to assess what other business units are doing, to avoid duplication of effort. It is possible, in an "organizational silos" environment, that one business unit might not know what another is doing, and consequently may duplicate projects.

Human Resources Considerations

The human resources consideration is often an essential factor. The organization must consider two aspects:

- Resource capacity and availability
- Expertise and competency

In some organizations, one or both of the above conditions may be in less than ideal supply, but the project is still of value and should be pursued. In that case, can the organization outsource successfully? What would be the impact on the project objectives?

Market Evaluations

Market research is often vital for certain projects, such as commercial projects releasing new products or services into the market. Market research can be extensive due to its direct impact on profitability or justification for the project. Market research includes analysis of market demand and supply (competition). One can understand that this is a very tough and subjective area, and because it is based on numerous assumptions, it might be a challenge to do it well. A potential pitfall, or risk, is being overly optimistic or pessimistic.

Social/Community Evaluations

Social or community evaluation is similar to market evaluation and is often necessary for social initiatives of non-government, non-profit, or government organizations. This type of assessment is necessary for these types of projects, and just because the project is not commercially driven does not mean the team can take short cuts.

The key questions are: Do society or community needs justify the project? What is required and what is feasible? How will the community use or operate the product of this project? Will the project depend on salaried, permanent staff or volunteers?[1]

> ### An Example of a Failed Project
>
> In a village next to the author's village, a political organization built a community center in 2010. However, they did not budget for staffing and operation. As of 2017, the building, unfortunately, is complete and unused.
>
> This is an example of a failed project and a waste of funds. The failure resulted from the lack of a full assessment of the project from all aspects, which includes operations.

---

[1] The author's first book about CAMMP™, *The Inheritance* (Ajam 2010), is a true story about establishing a non-profit organization that will build and operate a community center (Ajam 2010).

Location Factor Considerations

The location factor will apply if the project requires a physical location. For these projects, the location could make an immense difference, impacting many of the other factors listed here. The location factor could be crucial, because some projects might be feasible in a particular place and not viable in another. For example, the feasibility of real estate projects, petroleum facilities, landfills, and other sensitive projects are highly influenced by the location factor.

Technology Evaluations

The technology evaluation depends on the domain or type of project. These might be factors related to the technical or technological aspects that the team must consider during this assessment. This factor is vital when the project requires the use of new, possibly unproven technology, which would add to the challenge of the project. Therefore, one reason for this factor is for the project team to consider whether additional contingencies are required to deal with the "new technology."

Legal/Regulatory Considerations

Certain projects will have to comply with legal and regulatory requirements. For example, construction projects will have to comply with health, safety, and environmental regulations. Food-industry projects will have to comply with health and food safety guidelines. This consideration will also include things such as licensing and permitting.

Sustainability Considerations

Today, sustainability considerations are becoming the trend and the right thing to do. As outlined earlier in the book, sustainability today is about the triple bottom line of people, prosperity, and the planet. GPM Global adds into the mix the need to focus on product and process. In other words, sustainability is not limited to environmental matters but must also consider the impact of the project on society, workers, human rights, environment, and other such elements.

Time Considerations

During the feasibility study, there are not yet enough details to allow the PMT to produce a detailed schedule, and this is not required. Time considerations are mostly about the key dates for the project, such as any physical, community, or management constraints, along with the required deadlines. In addition, the team must assess if the targets or required dates, if any, are reasonable and achievable.

Cost Considerations

Cost consideration is similar to time consideration, and the team cannot produce an accurate estimate at this early stage of the project. At this juncture, the team can only develop rough, order-of-magnitude estimates. Such estimates could have a range as large as ±50% or more. The financial analysis would consider this wide range.

Other Considerations

It is important that practitioners not narrow their considerations on what is listed here. They need to think out of the box and about their project environment.

Major Project Risks

As outlined earlier, the feasibility study itself is a form of risk assessment, with a focus on the overall project and business risks and whether the project threats are too high to justify continuing with the project.

In addition to the question of whether to exploit the opportunity or not, the organization needs to consider the specific risks. During the feasibility study, the focus would be on risks related to the expected benefits (outcome) and the business risks, such as market conditions, competition, and market demand. However, the critical factor is to assess whether the risks (threats) will prevent the organization from realizing the anticipated benefits and achieving the expected outcome.

Risk does not have to be a separate topic in the study if the team embeds it in each of the above topics. However, we separate it here to emphasize it, and to indicate that these risks are different from the risks the project management team will manage during the Development and Delivery Phases. It is also worth noting that once management approves the project at SG2, they would by default have accepted some of these risks.

Financial Considerations

Financial considerations are left to the end, because they depend on all of the other factors. Financial considerations are about the project financing or funding and economic returns. That is, whether the project requires funds from outside the normal day-to-day operations—funds approved specifically to a project. Also, this would be the place to perform financial analyses, such as cash flow, profitability, break-even, return on investment, or any other financial factors that would help management decide the feasibility of the project.

Study Conclusion and Recommendations

The conclusion is simply the view of the project team on whether the project is feasible. A project might be feasible if certain conditions are met, so this would be a place to include such "if" conditions.

19.3.4 Feasibility Study Potential Pitfall

Due to the nature of most projects with significant ambiguity at this early project timing, the team will need to make many assumptions. Human nature has taught us that if an organization is favoring the project, its management could short-cut the process, amplify the positive aspects and assumptions, and trivialize the negative aspects. The opposite might also be the case: If certain executives do not want the project, they might try to bias the outcome toward an unfavorable conclusion.

If this is a reality in your world, then the feasibility study as a whole is questionable. Consequently, the organization would be taking a huge risk, which could result in a less-than-optimal solution, if not a failed project. Therefore, there has to be a culture of neutrality to remove or minimize any biases.

19.4 Stage Gate Two (SG2)

The feasibility study is an essential requirement within the project life cycle and would lead to Stage Gate 2 (SG2). This control point would provide management with the opportunity to validate that the team did perform a proper feasibility study. If management is satisfied with the level of effort and comfortable that the study is unbiased, then management would consider the result of the study and recommendations. *The vital question here: Is the project feasible?*

Satisfying the feasibility question leads to the second objective of SG2, which is to decide if the project is a priority and should proceed to the next stage. In other words, this is the time for the initial go/no-go decision.

If there is a go decision, the next step is for the sponsor to develop and issue the PAD, officially launching the project.

If the decision is not to proceed, the team will perform project closure, documenting the justification for stopping the project and any lessons learned. It is also possible that an organization could decide to hold a feasible project for future development.

> **Project Prioritization**
>
> If a project is feasible, management will have enough information to make a decision on whether to authorize the project. However, before they make the final decision, they should consider the project as part of a pool of other projects, a portfolio of projects and programs. In this case, management should approve the high-priority projects using whatever criteria they believe appropriate for their organization's strategic objectives. This concept is project prioritization, which is part of portfolio management.

19.5 Project Authorization Document

19.5.1 Purpose of the Project Authorization Document

The project authorization document (PAD) is the document that authorizes the whole project; it is what PMI and ISO label the *project charter*. What does authorization mean?

1. It announces to the organization that management has authorized a new project and is allowing the use of company resources *for the project*.
2. It identifies the project sponsor and project manager and their authorities.
3. It serves as a key reference and control document for future work.

Does this mean the project manager can proceed with the project all the way to completion? Usually, the proper answer would be "No."

What the PAD is telling the organization and the project manager is that executive management has a clear intent to take the project to completion, yet they are only *effectively* authorizing the next stage. The decision whether to continue with a project will be a subject with subsequent stage gates.

19.5.2 Content of the Project Authorization Document

How should the PAD look, and what are its components? It should be brief—ideally, a page or two; it builds on what has been done before and is the foundation for what is to come. The components are:

1. **Product (output) description.** The product description is a repeat from the project brief. It is a short, concise description of the product of the project. One sentence is enough, since this is not the detailed scope of work documentation.
2. **Project justification.** The project brief also includes project justification. Again, one sentence is enough.
3. **Location.** If applicable, specifies the location.
4. **Time and cost.** The PAD is not a schedule, nor is it a cost estimate. However, it includes the expected completion date and, possibly, other essential milestones. If the project has a time constraint or a window of opportunity, the PAD should include this condition. For costs, a rough order of magnitude budget would be required. This cost and schedule information would be extracted from the first estimate completed during the feasibility study, but management could alter the targets, if necessary.
5. **Project manager.** CAMMP™ calls for the project manager to be on board after SG1. If that did not happen, the project manager must be on board at this time. In either case, the PAD is announcing the project to the organization and letting the stakeholders know the identity and authority of the project manager.
6. **Assumptions and constraints.** At this time, it is important to list the major assumptions and constraints for the project.
7. **Major risks.** Risks that are not resolved during the Discovery Phase would be carried over into the Development Phase, if necessary, and monitored for further action.

> **Assumptions & Constraints**
>
> Assumptions are necessary as a by-product of the project development process and stage gate concept. Since it is not possible to know certain details about the project early on, it is necessary to assume certain things. The project manager and the team must address these assumptions as the project progresses. These assumptions will become risks if the team does not address them when required.
>
> Constraints are restrictions imposed on the team by management, clients, and nature, among other factors. These are the factors that limit the team in one way or another. The project manager's challenge is to deliver the project within the given constraints.

8. **Stakeholders.** The key stakeholders will be identified and listed in the PAD. However, a full stakeholder identification and analysis would be part of the next stage and updated with every stage after that.
9. **Project success.** If possible, the PAD should include the project success criteria, or at least key measures. In some cases, the organization cannot establish all the measures of success at this time, but they must be finalized during the next stage or the Strategy Stage.

19.6 Stage Summary

The focus of the stage is to understand and validate the idea for a new project, leading to project authorization. The team must conduct a proper feasibility study, and management has to accept that this idea is feasible and is a priority in comparison to other potential projects. It is critical for the organization to believe that pursuing this project will deliver the right outcome and realize the expected benefits.

Chapter 20

Development Phase – Requirements Stage

20.1 Phase Overview

The Development Phase is necessary for developing the concept of the authorized project to ready it for the Delivery Phase. In this phase, the level of effort is more than it was for the previous one but still low in comparison to the next phase.

CAMMP™ splits this phase into three stages: Requirements, Strategy, and Definition.

20.2 Stage Overview

The focus of the first stage, the Requirements Stage, is to define the project requirements—all the needs from all the relevant stakeholders. The stage concludes with the submittal of the project requirements document (PRD) for review and decisions at Stage Gate 3, which CAMMP™ labels the *Stakeholders Alignment* Gate.

20.3 Sequence of Events

Based on the above, the sequence for this stage is as follows.

1. The project authorization document (PAD) marks the start of this stage. Consequently, it is likely that there is no need for an independent stage authorization for authorizing this stage.
2. The project manager will assess whether additional project management team members are required to develop the stage management plan (SMP).

3. From the SMP, the project manager will identify and mobilize the technical/functional leads to develop the stage detailed plan (SDP).
4. The project manager will mobilize the rest of the team if additional resources or other specialties are needed to complete the stage work.
5. The project team reviews the PAD, identifies other sources of information, determines the stakeholders, and performs the necessary analysis to develop the project requirements document (PRD) based on the project needs and stakeholder expectations and requirements.
6. Next is Stage Gate 3 (SG3), which is the stakeholders' alignment gate. This gate is necessary to ensure the alignment of the project requirements and objectives among all relevant stakeholders, which should minimize changes later.
7. With an alignment at this gate, it is time for the project team to move to the next stage.

20.4 Project Requirements Document

20.4.1 Overview

The PAD is management, through the sponsor, communicating to the project manager and organization that the project is a go and represents the high-level perspective—what is required and why. The project team reviews the PAD, researches other organizational resources, and starts the work to determine the requirements for the project. At the completion of the PRD, it would be the project manager's turn to communicate with management and various stakeholders on the team's understanding of the project and what the team must accomplish to deliver the project's final output and outcome. In other words, it ensures that the project team, management, and stakeholders have alignment with the common goal and objectives.

20.4.2 Why the PRD?

This topic often leads to confusion on what the requirements are and why there is a need for a stage gate at the end of this stage.

The PRD is a vital part of developing the concept from the project brief, which is, by its nature, brief. The term *developing* refers to a further elaboration of the requirements in terms of output (product) and outcome (results, benefits).

The reader will notice that many of the topics of the PRD are similar to the PAD content, except that the PRD offers more detail than the PAD—it is the *progressive elaboration* as the project moves from one stage to another. Therefore, the project feasibility study and authorization document are essential input to this stage. In a later stage—the Definition Stage—the project team will expand on the PRD, through additional analysis and development, to produce the project detailed plan (PDP).

In other words, the PRD's level of detail is an expansion of the PAD, but not in enough detail to allow the implementation work; this is why we need the PDP, which has a significant amount of detail.

20.4.3 PRD Content

Figure 20.1, lists the various components of the PRD. This list is not exclusive since, in some domains, there might be more or fewer topics to consider. Organizational preferences also have an impact on the PRD structure and content.

- **Description of the product.** The first step is to describe the output of the project—the product. At this point, any additional information that the team develops or gathers will be included in the description. The team is not yet ready to produce the detailed scope of work; that is part of the PDP. Therefore, the need is to provide for enough detail to clarify the requirements and use the information later to develop the PDP.
- **Characteristics.** The thesaurus defines characteristic as "description, character, quality, feature." Therefore, the PRD description of the product will include required features, where applicable. For example, a house is a *product*. The features of the house are that it is *a three-story villa with huge windows and a large open space for entertainment*. Further features, or characteristics, are *a garden to include a children's pool and a Waterfall*.
- **Expectations.** The project team needs to be able to distinguish between *expectations* and *requirements*. Expectations are implicit, which, if not explored and identified properly, will likely lead to a lack of clarity, misunderstandings, delivering less than optimal product, and ultimately the possible dissatisfaction of the client. Therefore, the challenge for the project team is to work with the customer and all relevant stakeholders—internal

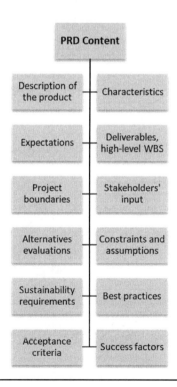

Figure 20.1 Content of the project requirements document.

or external—to *explore for the expectations*. Many tools and techniques can be used to explore for the deep-rooted expectations, but these are outside the scope of this work. Next, the project team must do their best to *transform those implicit expectations into specific requirements*. One possible approach to help the project team accomplish this task is asking lots of exploration questions of the client.

To clarify the previous paragraph, let us revisit the house example.

An implicit expectation might be, *"We have many friends (couples) who visit with us for vacation."* Transform this to a requirement: *"Suggest providing two guest rooms that are master suites for the privacy of the visitors."* Another example of an expectation: *"We should be able to take advantage of the great weather here year round."* Transform this to requirements: *"Suggest providing nice outdoor facilities for relaxation and enjoyment of the weather."*

- **Deliverables/high-level breakdown structure.** From the above characteristics and requirements identification, one can then develop a list of the key deliverables for the project.

  For the house example, the final deliverables (the output) are *a three-story house, four bedrooms for the family and two master suites for guests, ample space for entertainment indoor and outdoor, three-car garage, landscape with children's pool and Waterfall, etc.* Note that these are high-level deliverables related to the project's final product. With the major product deliverables and requirements known, the team can develop a high-level product breakdown structure (PBS); see Figure 20.2. As the project progresses, the list of deliverables is likely to grow significantly.

  At this point, the PBS does not need to have significant detail; possibly two to three levels only. As the team continues to develop the project, the PBS will grow in direct relation to the project size and level of detail. It would also be necessary to develop a work breakdown structure (WBS) to represent the project scope deliverables that will have to be produced and developed, leading to the final product *(see the textbox PBS or WBS)*.

- **Project boundaries.** Sometimes, one of the best ways to define something is to identify what it is not. In project management, that brings us to the closest thing to *project boundaries*, also referred to by some as *project framing*—in other words, what the project scope includes and what it does not include. Boundaries are vital to identify and are crucial for clarity of the project and its components.

- **Stakeholders' input.** When the team is working on converting expectations into requirements, it is vital to ensure communication among all relevant stakeholders. One common pitfall is missing the non-obvious stakeholders.

  (Refer to the textbox Appropriate Stakeholders for a story that demonstrates this point.)

> **PBS or WBS**
>
> "The WBS is a hierarchical decomposition of the total scope of work to be carried out by the project team to accomplish the project objectives and create the required deliverables" (*PMBOK® Guide* 2017, Part 1, Section 5.4, p. 157).
>
> Therefore, the WBS focuses on the project deliverables, while the PBS presents the breakdown of the product and its components.
>
> This is one of the areas of confusion in distinguishing between the product-*oriented* processes and the project *management* processes.

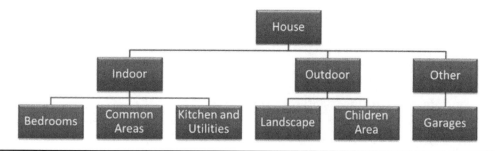

Figure 20.2 High-level PBS (house example).

- **Alternatives evaluations.** The evaluation of alternatives to the project, scope, schedule, location, and procurement approach would take place at various stages of the project, and in each stage, the alternatives can be different. Some of the alternatives could be due to the nature of the project: a house could be one, two, or more stories. These are alternatives to consider. Should the house be a standard house, a "green" house, or a "smart" house? During this stage, the team is likely to introduce value management and value improvement practices, and those will certainly lead to a few alternatives.

  Regardless of what alternatives the team considers, it would be best to address these alternatives during the development of the project requirements document. Ideally, the team and management should not leave significant alternatives open beyond the PRD. If there are unresolved issues and alternatives, the project would be at a higher level of uncertainty, which is a threat to project objectives. The story in an earlier chapter on the mega project on an island is a good reference here. On that project, the team had to choose between three implementation alternatives. Ideally, the team should have resolved that decision during the Requirements Stage, or during the next Strategy Stage, at the latest. The timing of such decisions is crucial because of the significant impact they could have on the project.

> **Appropriate Stakeholders**
>
> In the movie *The Towering Inferno*, directed by John Guillermin, Paul Newman is the architect who led the design work of the tower/skyscraper, and Steve McQueen is the captain of the fire fighters.
>
> During the opening ceremony for the tower, a small fire starts as a result of electrical problems (quality and cost-cutting issues). The fire quickly spreads, and the whole tower is in danger, along with the lives of the many people celebrating the opening. Consequently, many people lose their lives in the movie.
>
> At the end of the movie, McQueen turns to Newman and says, "One day you engineers are going to kill a thousand people in one of these things if you don't ask us how to build one."
>
> Obviously, the main point of telling the here story is about ensuring that we must get all stakeholders involved, even those who might not be within the performing organization, or that might not seem obvious. We have to think outside the box.

- **Constraints and assumptions.** The PRD must revisit the assumptions and constraints from the PAD to determine whether some constraints have been relaxed or removed, or new ones added. Furthermore, since the team is already progressing with the project, they should be able to clear some of the assumptions. However, new assumptions might come in as well.
- **Sustainability requirements.** What are the sustainability needs of the project? Would the organization consider all aspects of the triple bottom line of people–planet–prosperity or only consider environmental aspects? Would they be required to comply with certain guidelines? Would they participate in global reporting? Therefore, similar to the other topics in the PRD, the sustainability requirements section is an expanded version of the information included in the feasibility study.

    Once again, a good reference for this topic is the work of GPM® Global.
- **Best Practices.** Best practices and value-improving practices are often domain-specific practices. Therefore, they could apply to some projects but not others. In either case, this stage is an ideal time during the Development Phase to introduce the value-improving practices, because they will introduce possible alternatives with impact on cost and schedule.

    What are some of these practices?

    The author is familiar with the practices in the capital projects industry, which includes well-defined and documented best practices such as constructability, schedule optimization, zero incidents (safety), front-end planning, and project change management.[1]
- **Acceptance criteria.** The challenge to organizations and project teams is to separate the concept of *completion* from *acceptance* and from *success*. The issue here is that some organizations are settling on completion of work as the only criterion for acceptance.

    What happens if the team did not complete part of the scope? Would the organization accept the project product? If they do, could one say the project is a success, a failure, or challenged? This stage is the right time to consider these parameters, although teams should have addressed some of them in the previous stage, before project authorization, even if not in detail.
- **Success Factors.** One might consider a project a success from a project management perspective if the team delivers it within time and cost parameters. If management wants to elevate project management into strategic project management, then it is vital to consider and assess whether the project succeeded in delivering what it was intended to deliver. CAMMP™ proposes four dimensions of project success, which are the subject of Chapter 28.

> **Value Improving and Best Practices**
>
> Value Improving Practices (VIP) and Best Practices (BP) are similar concepts. They both deal with practices that are not commonly used by the average organization, and organizations that use them properly can expect significant performance improvement in the areas of safety, quality, cost, schedule, performance, etc.
>
> VIP/BP should be limited in number because they are often differentiating factors between typical performance and best-in-class performance.

---

[1] For further reading, a primary source would be the Construction Industry Institute (CII).

20.5 Stage Gate 3 (SG3)

20.5.1 Overview

With the PRD complete, this is an opportunity for executive management, the client, end users, operation and maintenance, project management, and other appropriate stakeholders to sit down again and review the work progress and the PRD. The main objective is for the stakeholders to reach an agreement on the project and to align on the expectations, requirements, and major deliverables.

For some types of projects, end users' representatives might be working with the project management team as an integrated team. In those situations, the alignment would be taking place on an ongoing basis; therefore, the organization would be able to expedite decisions at SG3. The decision, in that case, would just be a formality.

20.5.2 Importance of This Stage Gate

Unfortunately, it has been proven time and again that communication gaps exist; however, the lesson is often not learned. If organizations do not ensure that all sides are talking about the same thing, the risk is that the project management team could run in a direction that might not be the direction required by executive management. Therefore, it is necessary to define the requirements and key deliverables and agree to them before the team moves forward and only discovers the mistakes later. This clarification is a crucial point for major projects, because a significant amount of work and substantial cost would be required in the next stage. Therefore, it is vital to be sure of the scope, objectives, and other factors. It is crucial to have alignment of all relevant stakeholders—this is the core purpose of Stage Gate 3.

20.6 Stage Summary

The Requirements Stage is the third stage of the project life cycle of the CAMMP™ Model and the first stage in the Development Phase. In this stage, the project team develops (elaborates) on the project concept (project brief) to help ensure effective planning and successful delivery.

The focus for the stage is to identify the project requirements. Without the proper requirements,

---

Communication Gaps

One Scenario

A manager tells a team member to develop a marketing plan but does not provide any real documentation, charter, requirements definition . . . Three weeks later, the team member comes back with a list of pricing for advertisements and an advertisement schedule and agreement. Is this output acceptable?

Another Scenario

Executive management asks a team member to write a book about project management. What is the project? Is it writing only or publishing? A team member might proceed and start working on publishing options, since in his mind the intent of writing a book is to publish it. This might be true, but is that what executive management wants?

As you can see, we need to have much more information defined and agreed to before we can proceed. If no alignment or clarity exists, there is a high chance of wasting resources and effort.

especially requirements that the team derives from exploring the clients' needs and expectations, there is a high chance that the project management plan and the project detailed plan (deliverables of upcoming stages) will not reflect the organization's vision and aspirations. With a poor plan, the chance of project success will be significantly reduced, if not totally lost.

Finally, Stage Gate 3 is crucial and should be the point at which the organization agrees to *freeze the scope*. Because the project is still on paper and has a relatively small number of people involved, changes can be readily absorbed. However, as the work moves into the future stages, changes will become harder and more disruptive. In the Implementation Stage, changes could be the leading cause of project failures. Therefore, ensuring alignment of stakeholders should minimize, if not eliminate, change.

Freezing the scope and controlling change is vital for project success. In capital projects, project changes could be a leading cause of failure. Consequently, CII has developed a best practice in which organizations should consider a zero change policy.

Once again, project teams cannot afford to spend the proper time to define expectations and requirements and then ensure stakeholder alignment at Stage Gate 3.

Chapter 21

Development Phase – Strategy Stage

21.1 Stage Overview

The Strategy Stage is required to emphasize the need to develop the project management plan (PMP) as the key stage deliverable, which incorporates the delivery strategy, especially for projects with some level of complexity. In this context, *delivery* represents the upcoming Delivery Phase and *management* reflects *management and control*.

It is important to highlight at this point that the presented PMP is somewhat different than what ISO or PMI define as the project management plan. A better term would be *project delivery and management plan,* but since that might be confusing, CAMMP™ keeps the term *project management plan*, with the emphasis on managing (directing, leading).

As before, the output of this stage will lead to another gate, Stage Gate 4 (SG4). Before the stage gate, it will be necessary for the project management team (PMT) to have an updated "partial feasibility" with cost and schedule information that can help management decide whether to continue with the project.

21.2 Sequence of Events

Most of the work in this stage will be done by project management specialists. Based on the above, the sequence for this stage is as follows:

1. The project requirements document (PRD) is a key input for this stage.
2. Authorizing the stage could be the documentation of the previous stage gate or an independent stage authorization document.

3. As before, there will be a need for a stage management plan and a stage detailed plan. Alternatively, these could be combined based on the project's needs, size, or complexity.
4. The project team will review all of the project documentation produced to date and proceed with the development of the PMP.
5. Next would be SG4. In addition to approving the plan, this stage gate may include management approval for advance funds.

21.3 Project Planning

Time for a refresher on Chapter 15.

As explained before, there are two types of planning processes. One is about management, and the focus is on how to manage certain subjects or functions, such as how to manage scope or how to handle procurement. The other planning process is about detailed planning actions, such as developing the project and product scope of work, developing a cost estimate and schedule that would be accurate enough for control, defining the quality standards that would be applicable for the project, developing a detailed staffing plan with clear roles and responsibilities, among other things.

Therefore, the CAMMP™ Model offers two types of plan: the project management plan and a project detailed plan, which would be applicable at the project level. Each stage will have similar plans.

Here are a few key points:

- The PMP is the overall management approach for the **project** and all of its components (phases, stages, functions)—that is, how to manage the project leading to delivering the final product. Such a plan includes budgetary cost and semi-detailed schedules.
- The project detailed plan provides the details for the project (scope of work, schedule, cost estimate, staffing plan), which the team will use during the upcoming stages.
- The **project** management plan is the stage deliverable for this stage; the **project** detailed plan is the subject of the next stage.
- The bolding of the term *project* above emphasizes that these deliverables are for the whole project and not for a given stage.

21.4 Project Management Plan

21.4.1 Sections of the Project Management Plan

Per the CAMMP™ approach, the PMP has four distinct sections, and each section has many subtopics (see Figure 21.1). This plan is quite comprehensive and could be a substantial document for a large and complex project. It can also be quite simple for a small–simple project, with many sections taken from the organizational system and templates for common repetitive projects.

Although the team can summarize all of this information in one document, it would be logical and practical to consider completing the strategy sections first, then the rest of the sections will build on this section. Furthermore, it may even justify moving this first section to perform it immediately after project authorization, before the Requirements Stage. The placement of this strategy section of the PMP is a function of the project's size and complexity.

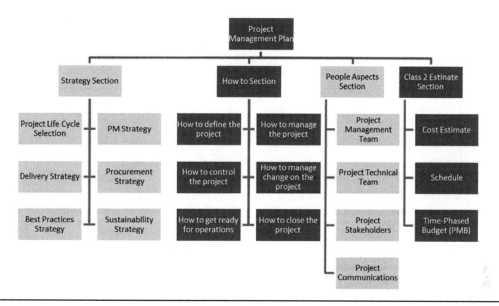

Figure 21.1 The project management plan with sections and subsections.

21.4.2 The Strategy Section

- **Project life cycle selection.** One of the first things that the PMP will address is verifying the proper project life cycle method to use. In a way, this relates to the strategy for managing the project—the general approach. Methodologies are customizable and adaptable. Therefore, within an organization, there could be more than one established project life cycle method that the project manager could use. The project life cycle selection is a function of the project classification—if the organization uses the concept of project classification.

  In addition to the pre-established standard models that might exist in the organization,[1] the project could have unique features that would justify a change to an established model, and the organization should empower project managers to *recommend* modifications to reflect the proper approach for a given project. The emphasis on *recommend* is to indicate that the project manager might not have the power to change the model but can recommend a change (with justification) to the project management office or OPMS (process) owner.

- **Project management strategy.** The project management strategy is about the general organizational approach for managing the project. It is a set of statements that establishes the general guidelines for managing the project. The use of the word *strategy* is to represent high-level thinking or approach, and not a detailed plan. These strategy elements are more like policies than procedures.

- **Delivery strategy.** The delivery strategy might not apply to many small–simple or medium–moderate projects, because they might not have any special circumstances that would require it. The delivery strategy is mostly related to the implementation of the

---

[1] Part of the organizational project management system.

work and the focus on the high-level thinking on how to do the job. The example of the modular construction of the mega project (in an earlier chapter) is a case of the delivery strategy. Another example might be answering this question, "Should we do the work using our resources or outsource it?"

- **Project procurement strategy.** Similar to the project management strategy, the procurement strategy is a set of guidelines about procuring goods and services to meet the project requirements. In some cases, the procurement strategy could have a major impact on the project development effort, hence the need to address procurement strategy at this time in the project lifecycle. For example, would the organization depend on local purchasing or global sourcing? local contractors or global contractors? Once again, the focus is on strategy and not on a detailed procurement plan.
- **Best practices strategy.** As stated in earlier chapters, guides such as those from PMI, IPMA, or ISO are about good or common practice. However, there is great value in implementing best practices on projects, which would help in enhancing performance and the chance of success. Best practices require a degree of domain experience, since best practices in capital investment projects are likely different than in software projects or pharmaceuticals projects. For example, Kanban or scrum might be considered best practices in software development, whereas constructability is a best practice for engineering and construction projects.

  Consequently, in this part of the strategy section, an experienced team can identify what best practices exist for their domain and decide if they can apply all of them, and if not, what could be applied effectively to the project. For capital projects, the Construction Industry Institute (CII) is the leading authority on best practices for this domain.
- **Sustainability strategy.** Similar to all of the other topics in this section, sustainability is still emerging, and this is why it is an advanced topic in the third dimension of CAMMP™. Along with all of the other advanced topics, the team shall consider sustainability at this stage. GPM and its PRiSM™ method include various aspects that can be incorporated here, first as a strategy and then in the "how-to" section below; it would be expanded as part of the PMP or an independent sustainability management plan.

21.4.3 The "How-To" Section[2]

Introduction

This how-to section includes the core components of the PMP. The team must focus on six core questions to complete this section:

1. How to define the project?
2. How to manage the project?
3. How to control the project?
4. How to manage change on the project?
5. How to get ready for operations?
6. How to close the project?

---

[2] In the context of the PMP, "how to" is inclusive of who, what, where, when, and how.

Obviously, the project team could add other questions, if necessary, depending on the project type and circumstances. Notice that four of these questions have a link to a Project Stage, and a couple of the questions are not limited to a stage. For example, *define* is related to the Definition Stage, *operations* to the Operational Readiness Stage, *manage* is linked to the Delivery Phase, in particular the Implementation Stage. On the other hand, *control* and *change* apply to all stages, although they are mostly active during the Delivery Phase.

How to Define the Project

Project definition is the focus of the next stage, in which the detailed planning for the project takes place, which leads to the project detailed plan (PDP).

Notice the use of the term *how to define* and not *define*. What is the significance of these words? When asked "how to define," one is only answering the how[3]! The answer might be, for example, "will need to survey our customers," or "will hire a landscape architect to design the garden." On the other hand, when asked to "define," this would be the actual design of the survey and the design of the garden, respectively.

The question here—how to define—is simplified and summarizes all of the questions necessary to finalize the definition package—for example, how to define the scope of the project; how to develop the schedule; how to manage risks; how to ensure proper human resources and communication planning; and how to define the requirements to meet the laws and regulations in term of health, safety, and environmental factors.

The best way that the team can properly and accurately define the project is by answering all of these questions and any others that might be more industry (domain) specific.

Remember: At this time, only answer the *how to*, not providing the detailed answers—yet![4] In other words, think before you act—or as some professionals like to say, "plan the plan."

How to Manage the Project

The previous point depends heavily on the project type and domain; therefore, it has to include quite a bit of detail, whereas this section could be very short if the organization has a proper project management organizational system. In that case, this section could be limited to "follow the organizational processes and methodology." In the absence of an organizational system, the project manager needs to explain his or her approach on how to manage the project to gain the acceptance of the sponsor and management.

As with the previous point, there could be many questions here. For example: "Who has the power and authority to approve x, y, or z?" and "How many and what positions are required for the project management team?"

---

[3] Remember, this includes the who, what, where, when, and how.
[4] This is a common area of confusion, especially with technical professionals—they would always want to put on their technical hats and jump into the definition, which must be avoided. Here, the team must keep on the project management hat, not the technical.

How to Control The Project

1. How to monitor project performance?
2. How to ensure proper project control?
3. What is project control?

Project control encompasses cost control, scope control, and time (schedule) control, among the other project management functions. A better term to use is *performance management;* this would be a direct indicator of project performance, which we check or measure by *comparison to the plan*.

Remember, control is happening from the idea and project brief. However, control in the earlier stages is mostly qualitative, but as the project moves into the Delivery Phase, control will be extensive and quantitative, hence the need to focus on "how to control" at this time.

How to Manage Change on the Project

The second essential component of control is the project's *change management*, which is specific to any proposed or approved changes from the plan.[5] Change management is like control—it should extend from the start of the project but become more focused in the upcoming stages.

How to Get Ready for Operations

First, let us define what is meant by operations.

Operations is the term used to refer to permanent operations of the product of the project, which is the responsibility of the end users—the client, internal or external. For example, if the project is to build a petroleum refinery, once construction work is complete, the project team will hand over the facility to the refinery management, who will be responsible for its operation and maintenance. For an IT infrastructure project for a facility, once the physical work is complete, the team will turn it over to the client, who will run the new hardware, operating and maintaining the new system—a call center, for example. The same concept applies for business or transformation projects. For example, the human resource department is the "operation agency" for a new employee performance appraisal system.

Therefore, the term *operation* refers to operation and maintenance, *after* the project is completed and handed over to the end users. In this context, CAMMP™ uses the term *operational readiness* to represent any work that is necessary so the client can operate, manage, and maintain the product of the project. As the name indicates, the term *readiness* is about all of the activities that the team must accomplish in advance of project completion and pre-handover—pre-SG6.

So, how to get ready for operation?

Typically, operation's personnel perform the work related to the operational readiness, with the support of the project manager and team. The main consideration at this stage is to decide how to perform this work. During—actually in parallel to—the Implementation Stage, the

---

[5] Notice the use of the word plan, and not project management plan or project detailed plan. This is by design, and will be covered in the project control chapter, Chapter 35.

project sponsor should form an operation team. This team would establish the necessary procedures, guidelines, training, and anything else that is needed to get ready for operation.

One final statement here: At this time, the project team, sponsor, and management should agree on whether a pilot (initial operations) period would be required. If necessary, the team must also address how to handle the pilot.

How to Close the Project

As the project moves into the Implementation Stage and progresses toward completion, a significant amount of work needs to start happening to deliver the project to the client (internal or external), and the team must also start planning for project close.

Some organizations do not put effort into project close—they view it as one of the least important project activities. There is no glory after the work is complete. However, from a strategic perspective,[6] the project work is not complete until the team formally closes the project.

21.4.4 People Aspects

Introduction

As the title of this section indicates, it is about the people aspects of project management, which consist of the teams, stakeholders, and communications. Note that for the earlier stage, this work would be part of the management plans for every stage. Here the focus is on the project, the remaining work, which is the Definition Stage and the Delivery Phase, with all of its four stages. For the upcoming stages, the PMP would be expanded to a higher level in each stage, as required.

The Project Management Team

The project management team (PMT) consists of the project manager and those other members whose job is to support the project manager in managing the project. These could include cost

---

[6] Remember, CAMMP™ takes a strategic, organizational perspective.

> **Mega Project Case Study**
>
> A new industrial facility is under development. A recent news item reported that the project (facilities) will be completed on a certain date, but commissioning and start-up will be delayed many months because the organization needs more time to get ready for operations.
>
> Remember, this is a mega project costing billions of dollars, and this delay will have negative consequences in terms of additional cost and delayed revenues and services, not to mention reputation.
>
> In a similar case study, a another major project has been delayed. A report came out saying the facility will be delivered this year but "we do not know when it will open."
>
> We do not have the full case studies, but based on the reports in the media, the leading cause of delay is in operational readiness, which implies lack of or inadequate planning.
>
> This is often a common challenge, because organizations focus on completing a facility and do not pay enough attention to what must happen next. Another challenge is that some practitioners do not consider this work as project management work.

and schedule specialists, project engineers, planners, and other resources. At this time, the PMP will include all of these resources, functions, numbers, timing on the project, and report relations. Therefore, the plan must include what is required for each stage, in terms of numbers, positions, and experience level. Figure 21.2 is a sample organization chart showing the various members of a project management team for a capital, medium, or large project, which includes engineering and construction of facilities.

A few clarifications:

- Depending on the size of the project, the number of PMT members could be much greater than shown or consist of only a few professionals. Furthermore, for small–simple projects, the project manager could be the one doing all project management work with no other support, even part-time support.
- For larger projects, the team could consist of a few members in the early stages and grow as the project progresses, peaking during the Delivery Phase, then reducing in number toward the end, leaving a few professionals to close the project.
- Q/HSE (quality, health, safety, and environmental) are shown here reporting to the project manager. In some situations, these positions would report to the home office with "dashed-line" relation to the project manager.
- There could be other positions that are not shown here, such as document control, communication manager, administrative support, technical engineers responsible for design supervision, and other possibilities.

The Extended Project Team

Per ISO and PMI, the project team includes three components: the project manager; the project management team, which we discussed in the previous paragraph; and the technical/functional

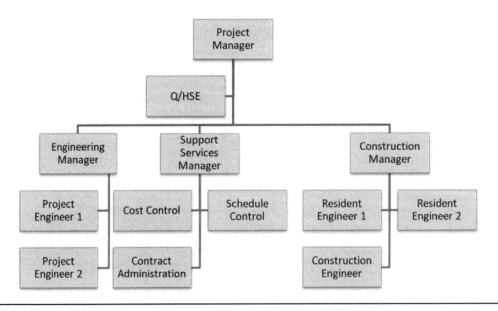

Figure 21.2 Project management team: a typical structure for capital projects.

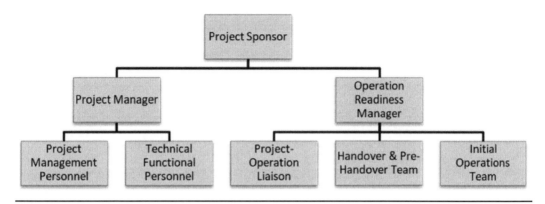

Figure 21.3 The extended project team.

team members who will perform the project work. These include designers, engineers, functional specialists, technical resources, and others, depending on the project domain.

CAMMP™ expands on this concept and considers a project team, in line with the earlier statements, and an extended project team, which would include the operation readiness team. Therefore, the extended project team consists of two major components, each led by a senior professional who would report to the project sponsor. Figure 21.3 represents this view and is a possible example of the extended project team that would be required to deliver the project.

The split under the operation readiness manager could be entirely different and a function of the project domain. Some of these personnel could be involved from the start and at every stage, while others will not be required until near the end of implementation. For new facilities, it is also likely that many will be recruited and trained to join the permanent operation team.

The extended project team will be updated and modified, as necessary, with each stage. It is even likely that most of the team members (technical/functional resources) will change from one stage to another. In other words, it is possible to have more than one team—for example, a feasibility team that is different from the definition team or delivery team.

Project Stakeholders

Similar to the project team, the project stakeholders could vary over the project's life, although some could be on the project from beginning to end. Therefore, the stakeholders' identification and analysis will be performed in the first stage and updated at every stage.

Project Communication

Organizations need people to manage projects; projects have an impact on people; and everyone affected has a need to receive or pass information. The PMP is the tool that the project manager will use to help establish the right mechanism for effective communication. Planning starts with identifying the communication requirement for the project's various stakeholders, then putting the necessary answers in terms of who communicates with whom, what information to share, how often, what format, and other relevant questions. The project team must be careful to address the needs of non-team members and all other stakeholders.

21.4.5 Class 2 Estimate Section

The term *class 2 estimate* will be clearer as we address estimating in Chapter 34. For now, let's think of this as the second estimate, with the first—the class 1 estimate—completed during the Feasibility Stage.

Cost and Time Estimate

During this stage, the project management team is likely to develop an updated time and cost estimate based on the new information the team has generated and defined since the feasibility study. The reader may have noticed that during the feasibility study was the first time to discuss the project cost and time. However, during the feasibility study, the team can only produce a rough order-of-magnitude estimate, because of the limited information available at that early time. With the requirements defined in the previous stage and project management planning work (including finalizing the project life cycle and various strategies) essentially complete, this would be the appropriate time to update the estimates and submit them with the PMP to management at SG4.

One common name for the estimate the PMT develops at this time is *budgetary estimate*, which is more accurate than the estimate used in the feasibility study but still not accurate enough for detailed control, and not suitable for final approval.

Performance Management Baseline and Time-Phased Budget

Once the project cost and time estimates and the high-level PBS and WBS are complete, the team can start to develop the initial time-phased budget reflecting the expected accomplishment (earning) of the project budget over time. This information is also useful for:

- Calculating inflation cost if the project lasts for an extended period—say, over six months or a year.
- Developing the cash-flow curves and expectations
- Calculating investment costs (loans or bonds) if required

In addition to the time-phased budget, the team can generate the performance management (measurement) baseline (PMB), which is the planned value curve, per the earned value management terminology. The PMB and time-phased budget included with the class 2 estimate are still considered preliminary, with the final cost and schedule information developed during the next stage, with the class 3 estimate.

21.5 Once Again: Process Groups

Some of the points in this chapter might resemble the planning process group of PMI and ISO. Therefore, at the risk of being too repetitive, it is important to revisit this subject.

Process groups are not stages of the project life cycle!

1. The PMP discussed here is for the project: the full project, not any stage or phase.

2. Furthermore, the guides from PMI and ISO refer to one plan—a project management plan (with subsidiary plans); whereas CAMMP™ splits the planning between a management plan and a detailed plan. Therefore, the plan discussed here is only the management aspects; the detailed aspects are part of the next stage.

21.6 Stage Gate 4 (SG4)

As a refresher, SG1 was about determining if the project is in alignment with the organization strategic plan and objectives. If so, then the organization considers the project for further evaluation and performs a feasibility study.

SG2 was about verifying that the project is viable (feasible) and whether management decides to proceed and authorize the project. A reminder: A project could be viable, yet management could still choose not to proceed with it because other projects might be of higher priority. This latter statement deals with the concept of portfolio management, which is outside the scope of this book.

SG3's purpose was on ensuring alignment among the various stakeholders on the project requirements.

At SG4, the main concern is for executive management to review the PMP and to determine if this plan would lead to the effective delivery of the project in line with the objectives that management has outlined earlier. Furthermore, management will review the updated cost and time estimates to determine if the project is still viable and still expects to realize the benefits.

Finally, management may need to approve a certain amount as an "advance fund." Advance funding is typically a financial and corporate governance requirement for capital-intensive projects. The team will need the advance funds for expenditures in the next stage, especially if the project team outsources this work.

21.7 Advance Funding

The advance funding point may raise two questions: how was the project funded up to this point, and why is no final funding available now?

21.7.1 How Was the Project Funded up to This Point?

At least in the capital project industry, and various other domains, the early work of the project—the initial stages up to this point—are funded from the normal operating budgets of the various organizational units involved. The organization's financial requirements might include documentation of these expenditures for later disposition. If management cancels a project during these stages, the funds are considered "sunk cost" and are treated as expenses in the financial statement of the organization. If the project is approved later, these funds will be transferred from the operating accounts and allocated to the project budget to be counted as a capital asset.

It is worth noting that on most projects, the expenditures up to this point might be less than 1% of the total expected cost for the project; therefore, the risk is acceptable. On the other hand, for small–simple projects funded from the annual operating budget, there is no need for the process described here; there would exist another process.

21.7.2 Why No Final Funding Now?

In response to why there is no final funding yet, the class 2 estimate is based on requirements and not detailed planning. Therefore, this estimate still has a high degree of uncertainty, which might not give enough confidence for management to provide final approval; it could be ±25% or more. Typically, final approval requires further definition and better accuracy to narrow the range and reduce risks. Most mature organizations, at least those working on capital investment projects, will not fund or provide final approval of a project unless the accuracy is ±10%, if not 5%.

21.8 Stage Summary

It is time to close this chapter with a recommendation.

The concept of planning is crucial to projects, and since CAMMP™ includes management plans and detailed plans at the project and stage levels, this can be quite confusing. Therefore, it is recommended to the reader to re-read this chapter after finishing Part E, or at least after reading the next chapter. Cross-referencing this chapter and the next one could be quite useful.

Chapter 22

Development Phase – Definition Stage

22.1 Stage Overview

The Definition Stage is the third and last stage in the Development Phase.[1] The word *definition* means different things to different people, and it is important to clarify this term.

- Some professionals think that the project definition is the high-level description of the project. In CAMMP™, this high-level description is part of the project brief.
- Other professionals think of project definition as the basic (preliminary) scope, which is the requirements document, discussed in Chapter 20.
- Others think of it as the project scope statement,[2] which is related, but not the same.

So what is the Definition Stage all about?

It is about the answers to the essential "how-to" questions from the previous chapter. It is about developing the project detailed plan (PDP). In addition to the PDP, in some cases, project funding (or approval) request is another key deliverable if the organizational system requires this to be a separate package from the PDP.

- There is only one stage gate, SG5, which is designed to review the PDP, and if all is well, management approves the plan and grants the final go/no-go decision.
- The PDP is an elaboration on the project requirements document (PRD) and also builds on the project management plan (PMP). It is mostly required to transform the needs,

---

[1] Other terms for the stage name could be *Preliminary Engineering Stage, Architectural Design Stage,* or *Front-End Engineering Design (FEED) Stage,* among other names.
[2] *Project scope statement* is a term used in the *PMBOK® Guide,* meaning an output of the Define Scope process (The Project Management Institute 2013a).

expectations, and requirements into clear deliverables. In addition, the PDP includes the definition of all other aspects, such as quality, resources, cost and schedule details, etc.

22.2 Sequence of Events

The main sequence in this stage will be similar to all other stages, as briefly summarized below.

1. Authorizing the stage would be done at SG4, using the stage gate documentation or an independent stage authorization document.
2. The project manager may need to mobilize additional resources to help develop the stage management plan, since the scope of the stage is substantially greater than before.
3. The project manager will need to mobilize additional technical lead resources to develop the stage detailed plan (SDP) if the PDP will be developed internally. If the work is outsourced, then the SDP could be elaborated by the service provider.
4. With the SDP completed, the definition team will move onto the project, and the full project team (internal or outsourced) will develop the PDP.
5. When the PDP is near completion, the project management team will work on the detailed cost and schedule estimate. Furthermore, financial or business planners will be working on the project funding (final approval) request.
6. Stage Gate 5 (SG5) is about the final approval of the project.

It is important to note that the cost of the work that the team performs at this point, from project brief to SG5, is minimal in comparison to the total project cost. On the other hand, the Implementation Stage is where most of the effort and associated costs will take place. Therefore, on most projects, this gate is classified as the point of no return. Once management grants the approval, it is likely that the project will proceed to completion.

22.3 Project Detailed Plan

22.3.1 Introduction

As described earlier, the project detailed plan (PDP) is partially about providing the scope definition and scope details. However, it will also include the detailed quality, health, safety, and environmental planning elements and the applicable standards, laws, and regulations. The detailed time and cost estimates are also part of the PDP. Furthermore, the PDP includes the details for all of the other project management functions and subjects. Notice the emphasis on "details," which means having all of the necessary information to enable ease of implementation.

In summary, this is all the detail the team needs to complete the project and deliver the product

> **Special Conditions About PDP**
>
> In certain industries, most of the work of this stage is very likely to be outsourced to a specialized service provider, such as an architectural firm for the architectural design or a technology company for an industrial facility.

successfully. It is about the work to be performed (the scope) along with how much it will cost (cost), how long it will take (schedule), what quality standards apply, among all other subjects.

22.3.2 Special Conditions

This stage is highly sensitive and directly linked to the type of project (industry and domain). Therefore, the technical content of the PDP (or whatever other industry-specific name is used) could vary widely from one type to another. The technical work required to deliver a front-end engineering design (FEED) package for an industrial or petroleum facility is entirely different from the work needed to provide the architectural design package for a major hospital project. The work is also significantly different for a talent development program versus an information technology infrastructure project or a marketing and outreach plan for release of a new product.

In line with the above, the PDP content discussed here is limited to the general project management requirements for the components to include in a PDP, regardless of the type of the project.

Where to start?

22.3.3 Scope of Work

- **Understanding the term *scope*.** In the project context, scope may be:
  - *High-level scope,* which refers to the project statement of work—the description in the project brief.
  - *Product scope,* which describes the features and functions that characterize the output—product. This output is what the team will deliver when the project is complete.
  - *Project scope* is the work that the project team needs to accomplish to deliver the output with the specified features and functions. This output will provide the capabilities to realize the benefits expected when the project was authorized.
  - *Stage scope* is similar to project scope, except it is specific to a given stage. For example, the scope of work of the Feasibility Stage is to deliver a well-prepared, balanced, and accurate feasibility study, for the particular project.

An example:

To build a house, the deliverable, in the end, is the house per the specified features; this is the product scope. The design, engineering, and construction are phases of the project, which are considered project scope. Completing the architectural design is part of the project scope and is the scope of a given stage.

What is not part of the scope?

The question of time, cost, risk, and quality are all project management factors that are essential to delivering the project and do have an impact on the product, but *they are not part of the scope.*

One might say the scope is to *deliver a house within the budget and on time.* That is not a proper statement and does not reflect scope; once again, the *scope is the house—only the house*—the physical aspects. Time and cost are requirements, constraints, related to delivering the scope, but are NOT scope. These are part of the processes of delivering the product. This concept is imperative to understand, since there are too many definitions for the word *scope*.

- **How to define the scope?** In the project management plan (PMP), one question was how to define the project, which includes the sub-question, how to define the scope?

  Possible answers would have been "will hire a consultant," "can do it internally with our team," "the team needs the help of other departments," as well as other options. The most generic answer would be that an individual or group—internal or external—will define the scope. In other words, with the PMP, the project team decided the how-to, and it is now the time to perform the detailed planning: to develop the project scope of work.

  In that regard, the scope definition starts with understanding the output or objective of the project, whether it is a physical product or a service project. Understanding the product and defining its characteristics was part of the project requirements document (PRD). Now, it is for the project team to act on their understandings and develop a detailed scope of work.

  The best approach is to start with the project requirements document and the initial, high-level PBS and WBS. From these references, the project team starts to elaborate and expand the product breakdown structure into the level of detail necessary to implement and control the project. The ultimate output is the detailed project scope of work.

- **WBS or PBS.** The term *work breakdown structure* (WBS), is often used loosely; it could mean various things, including a product breakdown structure (PBS). Distinguishing between the two terms is important. A PBS is a breakdown of the product and its components, whereas a WBS is a breakdown of the work required to deliver the product. The PBS represents product scope, and the WBS represents work scope.

  Back to the house example: A PBS would include the breakdown of the house components, as shown in an earlier chapter. On the other hand, the WBS would include (at level 2), things such as architectural design, detailed design, engineering design, and construction. At level 3, each of these major headers will have a further breakdown. For example, zooming in on construction, there will be site preparation, excavations, foundations, structure, electrical installations, and all other work.

  In closing, a PBS helps visualize the output, whereas a WBS clarify the journey—the work required to reach that output.

22.3.4 Q/HSE

- **Introduction.** Once the team defines the project and product scope, it is important to establish the quality standards that apply to the project, along with the health, safety, and environmental factors, if applicable. Remember, the question of how to define the project includes how to set the quality, health, safety, and environmental requirements for the project.

- **Why combine these four elements?** These topics (Q/HSE) are quite similar, from a project management process perspective.

  How are they similar? They all include planning, definition, assurance, and control steps. These topics are also usually related to compliance with laws or regulations. Although they are grouped here, in the PDP each could have its section, along with sustainability.

- **Define Q/HSE.** The team must establish the necessary plans that will guide them toward ensuring compliance with the standards and regulations.[3] Next, the team defines the Q/HSE standards and regulations that apply to the specific project or project domain.

It is worth noting that often, Q/HSE considerations are all highly linked to the organizational culture and management *stand* on the importance of these factors.

The team cannot limit their concern to "the work" during the Delivery Phase. Rather, it is essential that the project management team make sure that once the facility is operating, it is delivering quality products; that it will work in a safe manner, with attention to the health of the employees and community; and it is environmentally responsible.

Similarly, Q/HSE considerations shall focus on the management process and the project team, to ensure working in quality, safe, healthy, and environmentally conscious way.

These topics vary in importance greatly from one project domain to another. For example, in the petroleum, real estate development, power, and similar construction-related industries, Q/HSE considerations are crucial to the success of the project. On the other hand, for internal business projects, some of these functions might not be applicable or as important.

22.3.5 Schedule

What is next? The project management team develops the detailed schedule.

Just to visualize the difference between scheduling in this stage versus in the previous stage: In the previous stage, the focus is on schedule management, such as how to manage, update, and control the schedule, and how to handle change to the schedule. Also, in the previous stage, the project

---

> **A Difficult Situation**
>
> One of the most difficult tasks that I had to perform on a project as a young engineer was to determine the "cost of safety."
>
> I was working as the project control engineer on a mega project, and a senior executive wanted to know how much we were spending on safety-related activities and was that amount justified. My manager delegated that task to me, and with research and my manager's guidance, I was able to deliver it. The money was absolutely justified, but was it enough or too much? I am not sure if anyone can answer this question objectively.
>
> Although I have been fortunate that when I worked on construction projects, like the one mentioned above, I was working for a highly conscious and safety-minded company, in the construction industry, it is common to have fatalities on mega projects. Despite our safety-conscious culture, and the money that we were spending, we still had a few tragic accidents, the worst two being that one worker lost his legs and another lost his life.
>
> Even with these incidents, our records, compared to global industry average, was excellent. Unfortunately, not all companies subscribe to high Q/HSE principles.

---

[3] This plan would be an elaboration on what is included in the project management plan

management team (PMT) develops the class 2 estimate, which includes the schedule but not in enough detail for control. In this stage, it is important to have a relatively detailed schedule, which is well prepared to give executive management a level of comfort to approve the project and to use for project control purposes.

The schedule is NOT the plan,
and a bar chart is not a schedule!

- **Schedule versus plan.** The following statements are given to clarify common confusions.
    A project plan includes the various subjects that are the content of the PDP. The schedule is the logical relationship among the various activities represented by a network diagram, which is the representation of the project activities and their logical sequence. A bar chart, on the other hand, is a time-based representation of the schedule network diagram.
- **Schedule development steps.** The process to establish the schedule is simple and consists of a few steps, which would be similar to what ISO 21500 or the *PMBOK® Guide* define. There are also numerous resources with information about planning and scheduling with the necessary level of information, more so than this book can cover. Therefore, this book will not expand the coverage of this topic.

22.3.6 Cost Estimating

Cost estimating is one area which is often oversimplified or misunderstood, even on major projects. Estimating also varies from one industry to another. In some industries or organizations, they might perform one estimate during the project life span, whereas others might have as many as three or more estimates.[4] This chapter highlights the key principles.

How to Develop Proper Estimates?

We will not attempt to detail all project management functions, and we will not bore the reader with top-down, bottom-up, parametric, ratio, factoring, or index estimating.[5] Our focus is on the fundamentals, and once one knows how to handle the basics, there are many resources to enable the next level.

Let us take a strategic perspective!

Organizations that have been in business for a while should have records from similar past projects. If they did not maintain records, all they need to do is design a simple spreadsheet and start saving project information for the things that repeat. Whatever the business is, starting to build a database is as easy as a notepad and a pencil or a basic computer program. Once the organization builds a database, it is relatively easy to estimate; it would be a matter of counting

---

[4] For a discussion on how many estimates, refer to Part G.
[5] These are the names of estimating techniques.

widgets and multiplying the numbers by cost per widget.[6]

If the organization does not have a database, then the team choices would be guesswork, calling on vendors and service providers for quotations, or asking someone else—a third party, who might not be accountable—to estimate costs. None of these choices will help in developing good cost estimates.

For proper estimating, a WBS is a great tool to use as a starting point. A WBS provides details of the project scope and work packages. The details allow for easier estimating and more accurate estimates.

- **Importance of proper estimate.** Without a proper cost estimate and a realistic schedule, the team might be starting implementation by riding on the "cycle of doom." It is as if one is driving with the wrong map: the chances of getting lost is high. The result is a waste of time trying to discover where one is, and how to get back on the right road. In the end, one loses time and money, and fosters frustrations and potential conflict among the stakeholders.
- **How to develop the cost estimate?** As with scheduling, the starting point is the WBS—this provides the details of the scope of work in small enough work packages to facilitate the estimating of the project cost, working from the bottom up (bottom as from the work packages).

---

[6] We know this is an over-simplification, but the main point is that without good history and a database, estimating is just an exercise of guesswork or just "going with the flow," and we can guess what the cost of this will be at the end.

An Estimating Story

Sometime back, I was leading a cost estimating team for a mega petrochemical project. Management was insisting that we do our best to be stringent in our efforts and not to include extra allowances—no extra costs. Actually, they instructed us to have a "lean" estimate.

Since this was a mega project, the cost (early 1990s) was close to a billion USD. In such a project, there are numerous gadgets, such as electrical bulks, pipes, pumps, instruments, control system, etc.

One team member felt that management was pushing us not to be "lean," but "too lean," and he wanted to do something about it. In the details of the section that he was estimating, we needed 10 units of a particular expensive gadget. In the estimate, he put 20. Now on a mega project, as we reviewed the estimate, it was not easy to see what he hid in the details, but those extra ten items meant approximately $1 million! As I was reviewing his work, I caught this discrepancy, and we removed the extra cost.

A few months later, we were in the engineering office working on allocating our approved budget into control accounts (developing the time-phased budget), so we had to go into the details. I noticed that my colleague had done a similar thing in another area. He was persistent, and this gave us an extra million dollars. On a project of that size, a million is lost in the round-off. That project was completed below budget!

In project management, this is what we call fat, padding, or extra cost, and we consider it as unacceptable behavior, and we cannot allow it. Yet, as this story shows, it is easy to hide extra costs (or time allowance) in an estimate.

Once the team estimates the cost of the work packages, it will be possible to calculate the overall cost of labor and materials. Next, the team will have to establish the costs of all other components, such as supervision, management, organizational overhead, escalation (inflation) costs, financing costs (if applicable), and reserves. There could be special allowances, which would be a factor of the project's special circumstances.

22.3.7 Staffing

Human resource planning is about preparing for the staffing requirement of the project's upcoming stages: numbers, timing of when to join or leave the project, among other factors.

In the PMP, the team addressed most of the items related to the human resources requirements to manage the project, culminating in the project management team. Now, it is time to address the other required resources, if they are not outsourced.

Once again, the WBS could be a great starting point to help the team expand on the PMP and define the various people requirements for the different project components. On some projects, it is likely that the organization will outsource some, if not most, of the project delivery work. In that case, the service provider might perform the detailed staffing.

Just a reminder: Every stage has its project team, which could vary from one stage to another.

22.3.8 Risk Management

Risk management is covered in Chapter 37 in Part G.

22.3.9 Sustainability

Sustainability is covered in Chapter 30 in Part F.

22.3.10 Project Procurement

- **Common pitfalls.** Working with various clients and students of project management, a common misunderstanding is that when they think of procurement and contracts, they think of the Implementation Stage. Many organizations do procure services for the project implementation work; they could also procure services for the earlier stages. Organizations can outsource feasibility studies, project management planning, project definition, even operation readiness. Therefore, procurement can take place at any Project Stage.

  Another pitfall, or challenge, in some organizations, especially government, is that they link a project to contracts. In other words, when they think of the project, they always think a service provider is doing the work. As a result, their budgeting is linked to contract value, since they treat their internal work as overhead and zero cost. They also depend on the service provider to do most of the planning and control work.

- **Procurement and the detailed plan.** The PMP addresses the general procurement strategy for the project Definition, Implementation, and Operation Readiness Stages. As stated earlier, some procurement activities could have taken place as soon as the feasibility study. However, on most projects, the organization can perform most of the previous stages' work with internal resources. Therefore, procurement might not have been a consideration until detailed planning and the Definition Stage, or on industrial projects, during project management planning.

In preparing the procurement plan, the key questions are:

1. What work should the team outsource?
2. What is the most suitable type of contract for the project?
3. Is it better to choose one or multiple providers?
4. Does the team need the assistance of a project management company to help manage the project?
5. How would the team tender the project or segments of it? Through advertisement, pre-selected providers/vendors list, or another mechanism?
6. Who would prepare the procurement documents?
7. Does the team need legal expertise, and where do they get it?
8. What contracting standard would be appropriate?

22.3.11 Other Considerations for the Project Detailed Plan

For the remaining items to include in the PDP, they would follow a similar approach to the other discussions in this chapter. The detail planning would be an elaboration of the earlier work and what was included in the PRD. Therefore, there's no need to provide further details at this time.

Also, the team must identify whether there are any special considerations or items to address in the PDP that are not listed here. Because this chapter cannot cover all possibilities, practitioners have the responsibility to think about their circumstances.

22.4 Request for Final Approval

Either the PDP or independent documents include a project schedule and cost estimate, along with the details of the other project management functions. In most industries, the cost of the work for the earlier stages, before Stage Gate 5, is a small percent of the overall project cost. However, after SG5 the team will move into implementation, during which it requires most of the project funds and effort. Therefore, SG5 is crucial—it requires attention and may include a request for funds for final approval. In certain industries, this deliverable might be the responsibility of the sponsor, who would complete the application for final approval with information from the PDP.

With the budget approval package, it is typical that the sponsor team will revisit the project feasibility and financial analyses to determine of the project is still feasible. The output of this analysis will be input to Stage Gate 5.

22.5 Stage Gate 5 (SG5)

The title for Stage Gate 5 (SG5) is the final approval. This title is to reflect that in most industries, once the team produces the PDP, inclusive of the scope, time, cost, Q/HSE, and all the other subjects, then management has enough information to make this final decision. The reason for the use of the word "final" here is that the chance of canceling the project after this point is extremely low. Please note that organizations have canceled projects at this point, or after it, but not commonly, at least on capital-intensive projects.

What are the important points to mention at SG5?

1. Since this is the final decision, approval means the official funding or financing for the project.
2. For many projects, the decision is made by management. For major projects, the decision is likely at higher levels, such as the Board of Directors or the highest level of government for public-sector projects.
3. In either case, funding might take weeks, or even months, if it has not been pre-arranged.

22.6 Stage Summary

The focus for this stage is to perform detailed planning and complete the PDP, which consists of the scope statement; detailed schedule and costs estimate; quality, health, safety, and environment plans; risk management plan; risk response plan; among other components. With this stage, the team also prepares the request for the final project approval, since SG5 is the point of no return on the project.

This stage is crucial for the success of project delivery. If the PDP is deficient, then it is likely to have negative consequences during implementation or operations. The reader is reminded of the importance of a good work breakdown structure as a factor to enhance the chance of success. A well-defined WBS and PBS are essential to all of the other sections of the PDP.

One final note: For certain projects, such as software development, the team could incorporate some Agile concepts. Doing so might mean eliminating this stage and combining the planning and design work with implementation in sprints or increments. That would be perfectly acceptable for these types of projects, since CAMMP™ is an adaptive methodological approach and could blend in more than one technique or special method.

Chapter 23

Delivery Phase – Implementation Stage

23.1 Phase Overview

Figure 23.1 presents the Delivery Phase, which consists of four stages: Implementation, Operational Readiness, Initial Operations, and Close. Usually, operations personnel will lead the operations-related stages separately from the team working on implementation. The operation team(s) may even report directly to the sponsor (not through the project manager, as described in Figure 21.3 in Chapter 21).

As before, each stage has a stage gate, and in this phase, Stage Gate 6 is related to more than one stage. This stage gate requires input from the Implementation and Operational Readiness Stages.

One final note: Figure 23.1 shows that the Implementation and Operational Readiness Stages are in parallel and start at about the same time. This starting time is not always identical, because in a highly complex project, especially major organizational change initiatives, operational readiness activities could start much earlier than implementation. In those situations, operational readiness could start with the Requirements Stage. For projects with no significant or complex operational readiness requirements, this stage could start later than shown.

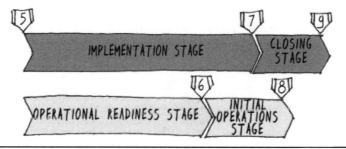

Figure 23.1 The Delivery Phase, per CAMMP™ Standard Model.

23.2 Stage Overview

The first stage in this phase is Implementation. This stage is usually the highest consumer of effort, time, and cost.

The focus of this stage is to implement the project per the project detailed plan (PDP). Implementation is the action of doing the "physical work" for the project to complete the required deliverables, leading to the handover of the completed product, service, or result (the output) to the customer, internal or external.

The time span of the stage is from final project approval at SG5 to handover at SG6. However, in most cases, some of the implementation work will stretch to SG7, and in rare occasions to SG8. In other situations, the implementation and operational readiness work could finish together, which would result in combining SG6 and SG7.

The major stage deliverables are the completed project work packages leading to the final output of the project.

Regarding the stage gates, SG6 is the first relevant stage gate. The purpose of this stage gate is for the teams (implementation and operation) to reach an agreement and gain confidence that the work has been completed satisfactorily to enable the handover from the implementation team to the operational side; this would likely be a provisional acceptance, not the final acceptance.

Next, SG7 is the stage gate that will officially recognize that the implementation work is complete, including all snag list items, and operations (representing the client or the sponsor) will issue a release letter. It is also possible to have the final acceptance certificate at this point, merging SG7 and SG8.

One last statement about this stage. For facilities projects, it is possible that this stage should be split into two or more stages. For example, in building projects, there will be a design stage (or detailed engineering) and a building stage (construction). Other types of projects might have similar conditions requiring a split of this stage into two or more stages.

23.3 Stage Sequence

The stage sequence is the same as in all other stages.

This stage will start with a stage authorization document, typically after final approval at SG5. Then the team will develop a stage management plan and a stage detailed plan (SDP), perform the work, and close the stage when done—all the while controlling throughout the stage. If the team outsources the work, then most of this work will be done by the service provider, but there will still be many activities for the project owner personnel.

In the case of outsourcing, it is likely that the service provider will complete the SDP.

23.4 Implementation and Control

23.4.1 Overview

The Implementation Stage is the stage that requires most of the effort, time, and cost. Most of the *project management* effort, at this stage, is about managing the day-to-day work to ensure delivery of the project per the requirements and the original project objectives. To achieve the higher objective—delivering the product—the project team delivers the pieces, which are called *work packages*. Furthermore, it is during this stage that the bulk of the project control effort is taking place.

Remember, control is not limited to control against the PDP during implementation. Control is happening from the project brief. However, since most control effort (the quantitative part) is going on in parallel with implementation, it would be good to address some aspects here, despite Chapter 35, which is dedicated to this subject.

23.4.2 Performance Management

Performance management is about how to manage the project performance throughout the project, especially at this stage. Some practitioners might prefer to use the term *performance measurement,* which is fine, but *management* is more results oriented in comparison to measurement. The team must establish the control reference points, or baselines, with a focus on the time-phased budget.

Let's start the explanation by going back to the previous stage.

In the Definition Stage, the project team completed the PDP, which includes:

1. The work breakdown structure (WBS), which provides the work packages, the scope deliverables, and the scope-of-work details. These items represent the *scope baseline.*
2. The detailed schedule, which shows the activities required to complete these deliverables, along with the project timeline. This detailed work is the *schedule baseline.*
3. The cost estimate, which includes the cost of these various packages. This cost estimate would be the *cost baseline.*
4. One can expand on the above and discuss quality baseline, staffing baseline, and other baselines that are in line with the project management functions.

The project manager and team can combine the approved WBS, schedule, and estimate and develop the project's performance management baseline (PMB). The PMB is indirectly a blend of the three baselines. One can develop the PMB from the project budget after allocating the various packages and control accounts over the project duration. This allocation would help create what is called *control accounts,* which would be key components for earned value management.[1]

The PMB is an essential part for project control; it is our navigation system and a key part of project management. Without it, one can think of the analogy of a plane flying without a navigation system—where it lands would depend on the circumstances.

Figure 23.2 shows an example of a time-phased budget and performance management baseline.

> The PMB is a time-phased budget plan against which project performance will be measured... The PMB should not include any management or contingency reserves, which are controlled budgets above the PMB.

---

[1] Earned value management is a performance management technique that is crucial for project control.

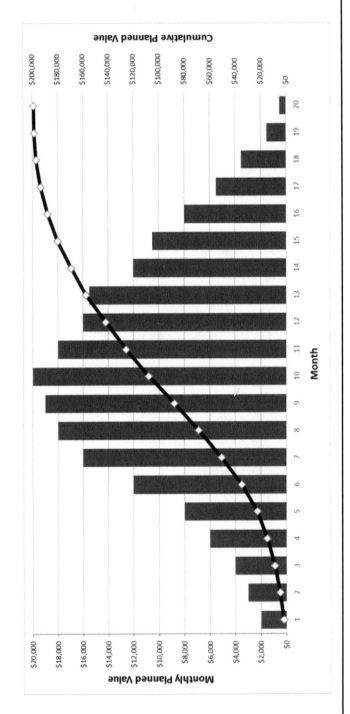

Figure 23.2 A sample time-phased budget and performance management baseline.

- The columns are the monthly planned values.[2]
- The figure also shows the performance management baseline, the line with dots.
- The PMB is the total curve, cumulative of the monthly planned values.[3]
- In earned value management, the PMB is also the budgeted cost of work scheduled (BCWS) curve.
- Typically, the team can develop this chart from the approved plan.

23.4.3 Start Implementation

The next step is to begin the physical implementation of this project. It is vital to re-emphasize that the word *implementation* represents carrying out the physical work and completing the project deliverables that the team defined in the PDP.

OK, clear, implementation is doing the work! But only the work?

Yes and no—control happens in parallel, as has been established.

As the team does the work, supposedly according to plan, we know that some things will not go according to plan; hence the need for humans to manage projects. Project managers can adjust as the work progresses, while keeping their eyes on the objectives.

Then what are some of the things that happen during the implementation?

It's time for a story!

23.4.4 The Trip (A Simple Project)

Project Brief to Project Detailed Plan

The story will start from the beginning. Let's take a road trip and consider the project as the trip; forgive the oversimplification, but one can relate to the concept and translate it to a given environment. We hope that your journey is not like the Abilene Paradox.[4]

This trip will follow the methodological steps that have been presented so far.

Once upon a time, a family wanted to go on vacation.

The idea. Take a road trip with family and visit two to three cities (*this is the what*). Purpose: have fun and relax (*this is the why*).

Approval (SG1). OK, a trip would be a nice vacation after this book is done (☺), and it is part of the family strategic objective, which is to explore different parts of the world.

---

[2] Refer to the scale on the left to read the value of each column.
[3] Refer to the scale on the right to read the value of each dot.
[4] "The Abilene paradox is a paradox in which a group of people collectively decide on a course of action that is counter to the preferences of any of the individuals in the group. It involves a common breakdown of group communication. Each member mistakenly believes that his or her own preferences are counter to the group's preferences and, therefore, does not raise objections. As a result, the group takes a trip, which they regret later." It is good reading. http://en.wikipedia.org/wiki/Abilene_paradox).

Feasibility. The cost for the road trip to a nearby country is less expensive than flying to a far location, and the family can afford it (*financial*). A good car is available to drive (*technology*). A good driver is also available, and there are enough electronics to entertain the kids between cities (*people*). There is no competition here (forgot about interruptions from the office to compete with fun—leave strong instructions not to be disturbed). You get the idea.

Stage Gate 2. The higher authorities approve the project; will prepare a PAD.

The PAD. The PAD is brief and includes the objective for the trip, where to go, approximate time and cost. For example, the PAD can include a trip to have fun and relax . . . will visit three cities in Lebanon . . . the assumptions are that the weather would be nice . . . constraints—pack light, we are in a car . . .

Project requirements. Have fun, activities for the whole family including children, the cities to see are Tyre, Tripoli, and Beirut[5] . . . emergency services for a vehicle, insurance, accommodations, cash and credit cards . . . based on the tourist information for these cities we should consider the following attractions . . .

Project management plan. The plan will include three cities with about five major attractions in each: fifteen attractions. On average, the family can enjoy two attractions per day, so it would mean about an eight-day trip, say ten allowing for driving and lazy days. The cost will likely be about $300 per day for accommodations, food, fuel, sightseeing. Determining quality requirements is mostly related to apartment hotels (since we are a family). Safety—practice safe driving. Health—ensure we have the proper clothes for the places to visit and include some basic medicines. Environment—trash bags for the car, so we do not throw it out of the window. People/family—two young boys, husband, and wife; most of the work will done by the parents, but will assign some actions to the kids. How to define the project? Tourist books for the attractions and maps for driving. How to manage and control? We will monitor expenditures and the maps, but the main objective of this trip is to have fun, so NO strict control.

Approval (SG4). The project is still in line with the original objective and expectations. Three thousand dollars overall cost is within the general budget.

Project detailed plan. The plan would include the details of the project, expanding on what we have done already:

1. **Scope.** Finalize the cities and define the attractions in each city. Prepare notes on what each city is famous for so we can purchase souvenirs. Mark the map for the road trip, with locations of all attractions.
2. **Quality.** Define the standard of the hotels and identify the hotels in each city.
3. **Safety.** Take emergency triangle for the car in case of breakdown; check the tires and engine; check security of the towns visited and what areas to avoid at night (or day, for that matter).
4. **Time.** Based on the scope, define how many days we need at each location and develop the schedule accordingly.
5. **Cost.** With the standard of the hotel defined, estimate the cost of the hotels, including any tax or service charge. Also estimate the cost of food and beverage. Calculate the distances to travel and estimate vehicle-related costs. Identify the cost of the attractions.

---

[5] Three cities in Lebanon.

OK, we think the idea is now clear; with a detailed project plan we would have completed the project definition. It is time for approval from the higher authority of the house.

The Rest of the Story

The final approval is in hand—time for implementation.

In this project, the Implementation Stage could be one stage or divided into three substages, with each city being a substage. It does not matter in this case, but it may on your project.

Packed and ready to go!

What could happen[6]?

Variances

During the implementation of any project, whether taking a trip or developing a green building, things will happen that would not be in accordance to plan. These are variances or deviations from the baseline.

What could these variances be?

The car is consuming more or less fuel than expected. Fuel cost along the road is more expensive than planned. The first attraction was great, and the family spent more time than planned there.

In other words, a variance could be cost, time, quality related, people related, or anything else that is different from what the team included in the PDP.

Some variances could be minor, and the team decides that they will be ok, no need to do anything about them. However, the team could also encounter substantial variances, which they should do something about to maintain a healthy project.

For example, let us say the family arrived at the hotel and found that it had raised its price by 30%, which is more than the family could handle. They can go to a lower-cost hotel, which would resolve this cost variance, but at a loss of time and some comfort! Alternatively, they can accept to pay the extra cost, with an impact on the project budget. The actions to correct a variance are called *corrective actions*.

Avoiding Variances

A project management fundamental principle is about taking preventive action and being proactive.

A good project manager should look ahead to avoid potential problems or variances. Listening to the traffic report indicates a potential traffic jam, so an alternate route might be a consideration. Check out the information about attractions to determine the best time of day to visit. These actions would help the family avoid long lines and reduce stress, leading to more enjoyment.

These kinds of actions—the ones the team takes to prevent a variance and enhance the chance of achieving objectives—are called *preventive actions*.

Rework

Rework is usually the result of work that the team did not do properly. It is also to repair a defect. Rework is mostly a quality variance.

---

[6] Notice that most of the items that will be discussed here are implementation and control processes.

For example, if the family encounters a problem with the car, they would have to fix it before they could continue.

Changes

The use of the word *change* here indicates any change[7] to plan.

What plan?

Since the project is in the Implementation Stage, the PDP is the main control reference point. However, a change could be to the project objective, as well. In this case, the change is in comparison to the project authorization document (PAD).

It is important to note the following:

1. Differentiate between a change and a variance.
   a. A variance is a deviation from the plan that relates to performing the work; then it is a performance-related matter, whether under the team control or not. *In the trip project, arriving late to Tyre is a performance-related variance.*
   b. On the other hand, project change is a conscious decision to alter the plan. *Deciding to go to Byblos instead of Tyre[8] is a modification of the project—a conscious decision.*
2. Differentiate between a change to the definition package and change to the project objective.
   a. A change to the definition package is possible while the project stays within the project objective. *During the trip, changing which attraction to see first is a change to the plan but remains within the project objective.*
   b. On the other hand, a change to the project objective is in comparison to the PAD. Such a change is likely to cause a change to the project plan as well. *Deciding to conduct some business while on the trip and visit with clients is a change in the objective.*

23.5 Stage vs. Process Group

What we've presented so far concerns the Implementation Stage. However, one cannot deny the similarities between the work of this stage and the implementing process group of ISO. The only difference is that the implementing process group represents implementing the work of a given stage, which varies from stage to stage. Implement process in the Feasibility Stage is to conduct the study and deliver the final study for management review. Implement process in the Definition Stage is to produce the PDP. Similarly, implement process in the Implementation Stage is completing the physical work for a given project, which could be design, construction, coding, writing procedures, conducting training—all depending on the nature of the project.

---

[7] Two points for clarifications: (1) change in the context of this book is about project change and not organizational change; (2) on construction projects, one might often use the term *variation* in lieu of *contract change*.

[8] These cities are a few thousand years old and have been continuously inhabited. They go back to the Phoenicians, who settled in the Eastern Mediterranean.

23.6 Operational Readiness

Depending on the type of project, sometime during the Implementation Stage, if not much earlier, the sponsor will need to assemble the operation team to start working on getting ready for the handover and the operation of the product of the project.

Handover refers to the project management team's handing over the responsibility of the output, the product of the project, to the operation group. Within this context, going through Stage Gate 6 will need input from two teams—the project implementation team and the operational readiness team. Both teams should have their corresponding work completed before the handover and before first operations. On some projects, both teams could be integrated and work as one team reporting to the project manager.

23.7 Stage Gate 6 (SG6)

It has been stated that lots of effort would be required to get ready for the handover and the operation of the *product of the project.*

The core purpose of SG6 is for executive management, the team, the client, and other relevant stakeholders to decide whether the project's work is complete enough to hand over to operation. The operation personnel must also verify this before they *accept custody* of the project's product from the project management team.

Please note that accepting custody is not final acceptance; it is only about changing the driver from project management to operation. It typically means that the physical work for the project is complete, and the client has accepted the work—conditionally.[9] However, the operations team still has to perform further testing and commissioning for final acceptance. In other words, this is a provisional acceptance.[10]

In closing this topic, it is important to stress that:

1. For some large projects or projects with multiple semi-independent packages, there could exist multiple review points, with acceptance for each component or package on its own, with the final gate approval happening at the handover of the last package.
2. The project management team is not out of the picture yet. There are always activities after the handover, such as supporting operation, fixing defects, and other work that is part of the project but is not essential to complete before handover.

23.8 Stage Gate 7 (SG7)

Stage Gate 7 is primarily related to completion of the implementation work. Once operational personnel and the implementation team agree that all snag items and minor defects have been fixed, and there is no more need for any implementation personnel, operations issues a release

---

[9] The use of the term *conditional acceptance* is for three reasons: (a) it is not the final acceptance, (b) the project team might still need to perform some non-operation critical work, and (c) there might still be some deficiencies that were not readily discovered during the handover.

[10] In industrial projects, this provisional acceptance might be called *mechanical completion.*

certificate indicating the completion of the implementation work. If there is no need for an initial operations period, then this stage gate will also indicate completion of the project work, except the formal project closure.

23.9 Stage Summary

Project implementation is a significant stage in terms of effort, time, and cost. It is often the phase during which the team is likely to face problems, especially if planning work was less than optimal. Alternatively, problems could be due to project circumstances that have changed drastically. It is during this stage that reality sets in and Murphy[11] often comes visiting.

The implementation effort and performance is a function of too many variables, such as the organization's maturity, the planning effort, the number and quality of the service providers, the project environmental conditions, along with external factors. The ability to manage people, resources, conflicts, requirements, expectations, and stakeholders can mean the difference between success and failure. Also, this is the stage during which the team captures anything that they could have missed—or face severe consequences during operation.

This would be an appropriate time to discuss project failure. Projects fail for numerous reasons, but one can categorize or group the reasons into one of the following four categories. Note that any of these categories could include individual or organizational factors.

- Poor or inadequate project set up and authorization; this could happen due to an improper feasibility study or not setting the project objectives properly.
- Poor or inadequate planning, whether due to team competence or lack of organizational system and historical data.
- Poor implementation; again, this could be due to team competence, organizational system, or other factors.
- External factors causing the project environment to change drastically but be beyond the control of the project manager and team.

Once again, implementation is the stage that would require most of the effort, consume most of the funds, and, possibly, during which one would notice a project going bad. The team must pay special attention to this stage.

It is also crucial that, during this stage, the project team and operation personnel work hand in hand to achieve a common objective. A statement easier said than done!

---

[11] In reference to Murphy's law, which says: "If anything can go wrong, it will."

Chapter 24

Delivery Phase – Operational Readiness Stage

24.1 Stage Overview

Every project that will result in a product and that requires operations will need an Operational Readiness Stage. Just to be clear, a *product* could be a facility, software, or an organizational system, whereas the term *operations* refers to the operational activities of the new system, including maintenance and upgrades.

The focus of this stage is on all the work that the teams have to complete before the handover and provisional acceptance. These activities include the development of procedures, training personnel, testing, and pre-commissioning. This stage is about anything that allows the starting of initial or full operations of the new product. In some situations, initial operations could be a pilot period after the handover. Initial operations is the subject of the next chapter.

Whether or not there is a pilot, and regardless of whether it is at this stage or as a separate stage, it is likely that such a project would include an operations period—after the handover, lasting until the client, operations, or end users, can grant final acceptance.

The time span of the stage is not well defined. It usually starts with either the commencement of the Implementation Stage, during the stage, or even before implementation. In the current model, the end of the stage is at handover, which would be the starting point for the Initial Operations Stage. In some projects, such as organizational change projects, this stage could start much earlier, perhaps as soon as the Requirements Stage, after the project authorization document (PAD).

There are many deliverables in this stage, which would be a function of the project type, but the key deliverable required from the stage gate would be provisional acceptance and handover.

As discussed in the previous chapter, SG6 is the key stage gate here, in which management, sponsor, implementation team, and operational team review the work completed and decide if they can exercise the transfer of custody via a formal handover from implementation to the operational team. The team accomplishes the custody transfer via a provisional acceptance certificate or a similar document.

24.2 Special Considerations

Before discussing the sequence, it is important to highlight a few points:

1. The Operational Readiness Stage might not be necessary for some industries, domains, or projects, such as some service projects or even research and development projects. However, on most, if not all, facilities, technology, or similar projects, this stage is essential.
2. The start and end of the stage are more dynamic than in the other stages; the starting point could be any time during project implementation, or in some cases even earlier. Although the end is provisional acceptance, where that is might not be clear.
3. It is common that the operations personnel, not the implementation team, will perform the operational readiness work. However, both teams must work in a highly collaborative manner. On some projects, both teams could work in an integrated manner under one manager. However, for facilities projects, the operational readiness team typically reports directly to the sponsor.
4. The author does not recall any project management book or guide from the various associations that includes operational readiness as an independent stage or even as a substantial topic. It is not clear why, although one can speculate two possibilities: (a) it is not required for every project, and (b) it is not typically the responsibility of the project manager. However, as with the Discovery Phase, CAMMP™ views this is an essential and substantial component of work that is required to deliver a project.

24.3 Stage Sequence

With the above clarified, what is the sequence of events?

Like any stage, it must be authorized, planned (management and details), implemented, controlled, and closed.

In terms of doing the work (implementing), the stage could be in two parts: the first part would be the completion of all the activities required pre-handover, the second would be the actual handover work. For some projects, the handover could be quite straightforward, and the team could complete it in a few minutes or hours. On the other hand, for facilities (industrial) projects, handover could take weeks or even months.

24.4 Pre-Handover Activities

The pre-handover activities could vary considerably from one project to another and one domain to another. In the process industry (oil and gas, petrochemical) and industrial projects, pre-handover activities can require a huge effort.

What are these activities? Again, they vary from industry to industry, but here are some suggestions:

1. For a new business,[1] the activities can include establishing the accounting systems and human resources policies, hiring, training, and marketing, among other items, as necessary—in other words, all of the activities needed to start operational work upon completion of the new business setup.
2. For a software implementation project, the activities can be preparing the testing scripts, procedures for change-over from the old system to the new, operations procedures, and training of staff on the new system.
3. For an industrial project, some of the deliverables would be operations procedures, safety procedures, pre-commissioning activities, commissioning procedures, staff training, maintenance procedures, the inclusion of the new facilities into the organization's change management system, and any other work that might be necessary.

Once the operational team accomplishes all of these activities and the work packages from the Implementation Stage are complete, operational and implementation teams would be ready to initiate the handover activities and transition the custody to operations.

24.5 Handover Activities

The handover activities are those that the teams perform around the time of handover. These activities include verifying work packages and product completion and verifying scope and product completion through inspections, testing, and reviews. To perform the handover reviews, the project management team, operations team, and service provider team will work together.

The objective of these reviews and inspections is to ensure that all items are complete per the project's requirements and the project detailed plan (PDP) and that they are ready for sign-off—acceptance—by the client. At this time, there might be some remaining issues that are not complete; these are "punch-list items," "snag-list items," or "exceptions lists."

If it is necessary to perform these items to meet the handover requirements, the implementation team must finish the work and verify completion before the client can accept custody. On the other hand, if these items do not affect initial operations, then operations personnel can proceed through the handover process and accept custody, on the condition that the implementation team will complete these remaining items post-handover and before final acceptance.

24.6 Stage Gate 6 (SG6)

The primary stage gate here is SG6, which was covered in the previous chapter. This stage gate would be in two steps: (a) ready for handover, and (b) provisional acceptance.

---

[1] Yes, launching a new business is a project. Notice the term *launching the business*, not the business itself. The project is complete when the business is ready to offer services.

24.7 Stage Summary

If this stage is required, then most of the work of this stage would be done by operational personnel or end users' representatives, with support of and in collaboration with the project management team.

Furthermore, most literature on project management does not discuss operational readiness, because this topic is often outside the responsibilities of the project manager. In some cases, these activities could be carried out by personnel from the home office of the organization.

Let us close with an example.

A hotel chain company is launching a new hotel. Usually, this would be thought of as a capital project (engineering and construction). However, it is much wider than this. What would be required for such a project? A team would be established to handle the facilities part such as requirements, architectural design, engineering design, and construction. Another team, possibly in the home office of the hotel company, will be working on hiring, recruiting, and training new staff, perhaps in another one of their hotels. Others will work on the accounting system, information technology system, security, marketing and sales, and any other tasks that are necessary for the handover and initial operations.

Remember, the Customizable and Adaptable Methodology for Managing Projects™ takes a holistic view of a project from the project owner's perspective. Therefore, CAMMP™ includes all of the activities that the team would perform in parallel to the design/build work. What CAMMP™ recommends is that a project manager manages all of these aspects with a deputy project manager for facilities work and a deputy project manager for operational readiness work. Alternatively, it is common today to have a sponsor with a project manager for the facilities work and an operational lead for the other activities. This is a matter of organizational choice.

Chapter 25

Delivery Phase – Initial Operations Stage

25.1 Stage Overview

As mentioned in the previous chapter, every project that will result in a product and requires operations will need an Operational Readiness Stage. It may also require an Initial Operations Stage. The term *initial operations* represents (a) a pilot period, (b) soft launch, (c) operations at minimum capacity, or (d) commissioning and start-up. The term is a function of the organizational preference and type of project.

The focus of this stage is to transition from work-in-progress mode to final status and operational mode. This transition period could last hours or days for simple projects and possibly weeks or months for industrial facilities or major real estate development projects. In some cases, it is possible that this stage would not be required, and the project would transition from the handover directly into full operations.

This stage could end with SG7, which indicates the end of all implementation activities. On the other hand, SG7 could represent the completion of implementation work, whereas final acceptance could be postponed to SG8, indicating that all of the project work is complete, except the project closure.

25.2 Stage Sequence

With the above clarified, what is the sequence of events?

As with any stage, it would need to be authorized, planned (management and details), implemented, controlled, and closed.

In terms of doing the work (implementing), the stage could be in one or two parts. For facilities projects, it is common to have two parts: Part 1 would be the completion of all start-up or commissioning activities, which is a form of final testing but is live testing with real products. Part 2 would be normal operations, but possibly at reduced capacity. Other types of projects might have a similar structure.

25.3 Pilot

The term *pilot* has been used more than once—so what does it mean?

Pilot is a term often used to mean experimental mode, simulation, sample, prototype, etc. The best way to explain this is with examples.

25.3.1 Example 1: A New Technology

Let's say a petrochemical company designed a new industrial process, but they have not used it before. The research and development process might initially require laboratory testing. If lab testing is successful, the organization might still want a higher level of confidence before investing millions or hundreds of millions in a full-scale project. In that case, the second R&D step would be to build a scale model. This is the pilot. If all goes well, they proceed with the full-scale installation. In this case, the pilot could be an independent R&D project, or a phase in the new facility's project life cycle.

25.3.2 Example 2: A Hotel Project

Let's say a hotel company wants to launch a new chain of hotels based on a new concept. They could build the new design in one location and monitor the situation for a while. They would collect feedback from the staff, visitors, guests, and others, learning what is working, what is not working, what is not optimal, and what is great. Then they take the learning from this first location, the pilot, and apply it to all new designs as they implement the new-concept hotel on other sites. The same can apply to other service-industry projects such as restaurants, hospitals, or retail outlets.

25.3.3 Example 3: An Organizational Change Project

In large organizations—conglomerates or multinationals—an organizational change project could be a huge disruption, and failure could be catastrophic.

This change project could be for a new technology or new business process, such as a new performance appraisal system. The organization could design the new system and implement it in one department, division, or location—a pilot implementation. Then, as in the previous example, the team would monitor the situation to capture the learning and update the system by fixing any errors and closing any gaps. Once the pilot is complete, the organization can move ahead with implementation in the other divisions and locations.

25.4 Initial Operations

The term *initial operations* is used to stress that this is not the permanent status, meaning no permanent or full-scale operation has been achieved yet. What do initial operations include? It all depends on the industry and the projects.

Commonly, the initial operation starts after the client accepts the output at the handover point, SG6. As mentioned earlier, at SG6, there is provisional or interim acceptance. It is also the point at which the project team transfers the custody of the facility to the client's operations team, which performs various commissioning, or start-up, activities in final preparation for the operational activities.

Once the operations team is ready and satisfied that all individual components of the products are okay and functioning as expected, they start formal operation of the facility, which is defined here as the initial operation. Initial operation and normal operation are essentially the same in terms of the work, or what is happening physically; the only difference is that the initial operation is taking place before final acceptance and possibly at reduced capacity. Operational personnel would closely monitor the initial operation, looking for potential issues. This period is necessary to build a level of comfort that all is working well and as expected.

The above is the possible situation in various project types, especially industrial and facilities projects. As mentioned before, for other projects, the initial operation might not be required.

25.5 Stage Gate 8 (SG8)

SG6 is the starting point for this stage—the point of handover and provisional acceptance. During this stage, SG7 could take place to indicate the completion of all implementation work, including any snag list items. However, the key gate here is SG8, which indicates final acceptance. At this stage gate, management will issue a final acceptance certificate stating that all of the work to deliver the project is complete, and the product is in line with the project detailed plan (PDP) and the objectives established at authorization.

25.6 Stage Summary

If this stage is required, most of its work will done by operational personnel or end users' representatives, with the support of and in collaboration with the project management team. The steps are clear and no need no further elaboration at this point.

Chapter 26
Delivery Phase – Close Stage

26.1 Stage Overview

The Close Stage is the last stage of the standard project life cycle of the Customizable and Adaptable Methodology for Managing Projects™.

The focus of this stage is on properly closing the project and includes all of the related activities, such as gathering the final lessons learned, updating employee files, documenting project performance metrics, updating organizational records, etc. During the Close Stage, the organization should also include project success evaluation and benefits realization, where possible.

The time span of this stage starts with SG7, at the end of the Implementation Stage. However, the final closure cannot take place until after final acceptance at SG8.

The major deliverable is the project close-out report. Although there are other items at this stage, they could be either part of the close-out report or independent deliverables.

There is only one stage gate applicable to this stage, Stage Gate 9 (SG9), which represents the organization management's acceptance of the project—all of the work, including the close-out activities.

26.2 Stage Sequence

The Close Stage is a stage like all others. Therefore, the processes apply to this final stage. The only exception is that closing this stage coincides with the project closure.

26.3 Stage vs. Process Group

Because of the similarities of the actions at this stage to the closing process group of PMI and ISO, it is vital to discuss the differences at this time.

All process groups repeat per each project phase or stage, and the process *groups* are not project *phases*. Consequently, every stage of the project has to be closed, following the appropriate Close process. Closing a stage would take place at the end of every stage, once the client accepts the stage work, typically at a given stage gate. Stage closure includes all of the deliverables that the team produces for the project, except that the focus would be limited to the given stage.

In that case, what is left?

At the Close Stage, the team will collect and summarize all stages' closure output as part of the final project close-out report. The final report could just be a collection of the stage reports, but it would be good to include other things that the team did not cover in a given stage.

26.4 Importance of Proper Project Closure

It is time to re-stress the importance of this stage and the close-out report.

As mentioned earlier in this book, some organizations neglect this stage, partially because they reassign the project manager and team almost as soon as the project is handed over. In these organizations, project completion coincides with the end of the implementation work or acceptance of the product.

On the other hand, CAMMP™ views project closure as an essential component of project delivery and a necessary enabling factor for organizations to achieve a higher level of organizational project management maturity. No organization can achieve this higher level of maturity without learning from the past and pursuing continual improvements. Continual improvement requires proper project closure, inclusive of lessons learned and update of the organizational records.

Keep in mind that significant effort and work takes place between the idea and project completion, which is crucial for organizational learning. *If project teams do not document the learning in a ready-to-access manner, a wealth of information is totally lost.*

Capturing the learning should not happen at the end only—it should happen throughout the project. One possible approach to avoid missing critical information is to assign a team member from the very start of the project to stay behind and close the project. Such a team member would be involved from the outset and at every stage.

Project Closure Story

Once upon a time, a long time ago, the author was mandated to establish a project support and control office to handle project control on an Alliance (multi-projects) contract. Under that contract, there was a portfolio of projects valued at more than $100 million (in the mid 1990s). These were projects in the petrochemical industry.

We established the office, formed the team, and as the team was starting their work, we discovered that the vast majority of projects from the prior five years were not closed. As a result, we had to assign a full-time team member just to close these projects.

That job was a nightmare, because most of the people involved were not available; so closure ended up being limited to an accounting exercise to ensure that we captured all the costs for the company accounting system.

Unfortunately, so much was lost that there was no proper reconciliation or lessons learned!

The CAMMP™ approach advocates proper closure by including the Close Stage as part of the Delivery Phase. Another factor is the introduction of a stage gate (SG9). These actions clearly state that the project is delivered to the owner organization only after it is closed and passes through this final gate. Without going through this gate, the project is not complete.

26.5 The Close-Out Report

What are the major components of the close-out report? It depends on the organizational policies. In the CAMMP™ Model, the final report consists of the following topics:

- Reconciliation
- Lessons learned
- Project success assessment
- Benefits realization evaluation
- Organizational records update

These elements could be independent reports or part of one document, a final close-out report.

26.5.1 Reconciliation

The term *reconciliation* refers to the final project performance reconciliation report, which is an essential document for organizational learning. This report documents what happened throughout the project from start to finish, with an emphasis on reconciling the differences between plan and actual.

Like the rest of the project closure processes, the performance reconciliation is an exercise for every stage and for the project as a whole. For example:

1. Within every stage, reconciliation is about the comparison between the final stage results and the stage detailed plan (SDP).
2. Along the project life cycle, reconciliation is about comparing each deliverable of a given stage with the stage entry point—the deliverable from the previous stage. At the end of the project, the comparison is between final state and the project detailed plan (PDP).
3. These reconciliation reports are also required for the assessment of stage and project success.

What are these reconciliations reports all about?

They are about documenting the differences between what the team planned and what actually happened on the project. These would be the performance variances or deviations.[1] The project team should also include the changes[2] to plan, reasons for the changes, and sources of change, along with any predefined project change metrics. This is vital in every stage for organizational learning.

---

[1] Variance is a performance deviation from the plan, possibly for external factors, but it could also be due to internal factors.

[2] The term *change to plan* refers to the conscious decision by the project stakeholders to change the plan or objectives.

To facilitate the reconciliation, it is best if the team focuses on the differences in each project management function.

1. **Scope.** The scope reconciliation documents whether the project team did deliver all of the scope required—no more, no less. If not delivered, then what is different and what are the reasons for the differences? Are these differences documented as changes or not, and if not, why not? What are the consequences of the missing or additional scope?
2. **Quality, health, safety, and environment (Q/HSE).** In this area, some questions would be: How did the team perform on these functions? How does the project safety record compare to the industry average or best in class? Did the project experience a higher rework ratio than the industry or was it in line with the industry? Did the team encounter environmental issues? What caused them and how were they treated?
3. **Time and cost.** How did the team perform in comparison to the project time and cost baselines? Is the final state within acceptable tolerances? Did the team need to use the reserve time and contingency allowances? Were they enough? Were the differences due to unusual circumstances or related to the planning tools? What are the factors that the team should communicate to the owner of the organizational records that would lead to updating these records?
4. **Human resources, stakeholders, and communications.** Did the team have any exceptional performances that the project manager should note and offer commendations for? How about the effectiveness of the communication plan—were there any gaps, and what went well or did not go well? Were the project and functional managers able to cooperate properly, or where there issues to address for the future? How did the team manage the relations with the stakeholders and engage them? Any lessons learned to share?
5. **Risk management.** Did the team follow the process well? How effective was the overall risk management effort? How about responding to risks? Did the team do an excellent job of identifying risks, or did they end up encountering many new issues during the Implementation Stage and execution processes of the various stages? Were the risks' assigned owners effective in managing the risks or were there issues that should have been addressed as learning opportunities?
6. **Contract management.** Contract management includes all of these topics but specific to contract work, so one can ask similar questions. Also, what are specific deviations or issues unique to the procurement and contract management processes?
7. **Summary of metrics.** Every project has performance metrics,[3] and this report should include a summary of these metrics, such as cost performance index, schedule performance index, project management index, change order index, quality index, etc.

26.5.2 Lessons Learned

Lessons learned are crucial, and the project management team must capture them throughout the various Project Stages.

---

[3] Performance metrics or indicators, like key performance indicators (KPI). These are limited to project performance and not the organization's performance.

How to capture lessons learned?

With a mature project management culture within an organization and the right mindset, every team member must be aware of the importance of lessons learned. Therefore, any team member who discovers a new lesson should document it. One must not wait until the end of the stage or, even worse, the end of the project.

At the end of a stage or project, the team should come together and reflect back to identify any other lessons that they might have encountered but did not capture at the time. Furthermore, as the team assembles the performance report, other lessons learned might appear. An experienced team (and mature organization) recognizes that lessons are about the good, the bad, and sometimes the ugly.

Final note: lessons learned can only work in mature organizations, because lessons should be clear and open; in a culture of blame or scapegoats, one cannot honestly expect many valuable lessons.

26.5.3 Project Success Assessment

Measuring project success is another way that the organization can evaluate whether the project delivered the anticipated benefits. The only difference from current practices or a narrow perspective that focuses on cost and schedule as the only criteria for measuring success is that, CAMMP™ includes four distinct dimensions, which are the subject of a Chapter 28.

At this time, it is worth noting that of these four dimensions, only two—product technical success and project management success—are possible to assess before closing the project, and the results must be documented in the close-out report. The other two dimensions, project delivery success and project objectives success, require a period of operation after project completion. In this case, the close-out report should include the dates at which these last two assessments would be possible, and the PMO or organization management would be responsible for performing these evaluations in due time—likely post project closure.

26.5.4 Benefits Realization Evaluation

An organization authorizes a project because executive management recognizes that the output of the project would add capabilities to the organization, which in turn, would produce the outcome and benefits desired.

Therefore, some of the questions here would be: Did the project add value? Did the organization realize the benefits that they expected when they authorized the project? Can the team even assess the situation at the time of closure? Benefits realization is a topic that is related to one of the dimensions of project success, which the organization can only assess in the future. On other projects, it might be possible to assess success and determine the realization of benefits with project closure, but these projects would not be the usual type.

A few more comments.

Let's consider the following: At the start of the project, the feasibility study determines if the idea is valid and viable, and whether the organization can deliver a successful project. At the end of the project, before or after closure, a benefits realization study determines whether the team achieved the desired objectives. In other words, the later study validates the earlier study. Without these studies, it is likely that organizations would not be practicing proper governance and asset management.

26.5.5 Organizational Records Update

Every organization that has adopted project management as a formal function will have an organizational project management system (OPMS) in place that includes the organizational records, which include data from completed projects, lessons learned, and reconciliation reports. All of these would be useful as a new team works on a new project.

One can understand that some organizations are quite young or are new to project management, and do not have a history; therefore, they might not have kept their project records in an organized manner for ready access. In either case, it is necessary to establish a system for record keeping and maintaining information from past projects. The records are likely to include a methodology, procedures, guidelines, policies, and other relevant information. Also, the records contain information on past projects, such as cost and time data.

Therefore, the team must incorporate the various closure reports into the organizational records. Furthermore, the project management team can summarize key factors from these reports to include, in a summary format, in a database for quick retrieval in the future and overall analysis of the organizational portfolio performance.

Let's refer to a hotel example again.

A well-established hotel company, with a proper OPMS and historical records, should have a project management information system as part of the OPMS. These records can include things such as cost per square meter of building a three-star hotel versus a five-star hotel in Dubai, plus approximate durations to build a four-star hotel or adjacent banquet facilities. They would have information indicating how many staff they will need based on the expected capacity or number of rooms/beds and a breakdown of the staff between security, front desk, restaurants, and all other facilities. If one zooms back on construction costs, they would have an idea on how much it costs for carpeting versus decorations. Without this type of information from previous projects, the hotel company would be re-inventing the wheel with every new project or depending on third-party estimates, which might not be available during the Discovery Phase.

26.6 Stage Gate 9 (SG9)

This final gate is quite different from all of the other gates, because there is nothing after it. The previous stage gates were important as control points. Most of them had two objectives: (1) accept the work of the previous stage or deliverables, and (2) if the work is acceptable, decide on whether the project is to continue on to the next stage. On the other hand, Stage Gate 9 (SG9) is only required to verify that the project closure is completed in a proper manner.

Because there is nothing beyond this point, project management literature does not address the need for a stage gate at the end, and the project is just closed. Some would consider the act of closure as a gate, but this is questionable if there is no authority verifying that the project closure is proper—a potential gap.

CAMMP™ emphasizes the need for this stage gate as a significant contributor to project management maturity, stressing the point that the project is not closed until management says it is closed.

26.7 Stage Summary

The Close Stage is the last stage of the project life cycle and would be the final chapter on the first dimension of the CAMMP™ three-dimensional model.

It is still somewhat common that some organizations either completely ignore this stage, or just close the project with a brief report that is of limited use and value. The CAMMP™ Model embraces strategic thinking, which means that proper project closure is essential for any organization that delivers projects.

With the strategic perspective, the focus of this stage is on properly closing the project, with all of the related activities. These activity reports have been defined as the project performance reconciliation, lessons learned, benefits realization study, success measurement, and updates of organizational records.

The final gate is necessary to stress that the organization does not consider the project closed until the team delivers a proper close-out report, and this gate will help ensure this status.

With this chapter, the discussion of the CAMMP™ first dimension is complete, stage by stage. Part D addressed the second dimension, which focuses on the project management processes and functions. Now it is time for Part F, which addresses the third dimension.

Section II, Part F:
The Third Dimension (Advanced Topics)

Chapter 27

Introduction to the Third Dimension

27.1 A Refresher

The first dimension of The Customizable and Adaptable Methodology for Managing Projects™ (CAMMP)™ provides a standard project life cycle model from idea to closure. This project life cycle has phases, stages, stage deliverables, and stage gates. The number of phases is set, but all of the others are variables in which the number of stages, and associated deliverables and gates, could be greater or fewer, based on the needs of a given project type.

The second dimension offers the project processes and functions that apply to every phase or stage. The foundations for this dimension were the PDCA cycle and the guides published by ISO and PMI, but as presented in Part D, CAMMP™ offers a significantly revised approach, with six processes and 12 functions.

One of the obvious gaps in practice today is limiting the project management practice to the second dimension only. This practice might be acceptable for small projects, but this dimension is far from enough to manage projects end to end and effectively. Therefore, any organization wanting to deliver value through projects must consider both the first and second dimensions, at least.

If the organizational project management system (OPMS) incorporates these two dimensions, it will enable good performance, which might be satisfactory for most organizations, or at least organizations with limited resources. However, organizations seeking excellence must view project management as a strategic imperative and consider project management skills as performance differentiators.

A journey to excellence requires a higher level of project management maturity, and an enabler for high performance and leadership requires the third dimension.

27.2 The Third Dimension

The third dimension is the concept added as a dedicated dimension in this third version of CAMMP™. (Version 2 included many of these concepts, but not explicitly stated.)

The best way to view this dimension is to think of it as a set of layers added on top of the first two dimensions. Each of these layers represents a set of practices that organizations would integrate into every applicable part of the first two dimensions, both along the project life cycle and in each of the stages.

It is not vital to implement all of the layers in one initiative. Organizations that have reached a level of comfort with the first two dimensions can start considering these additional layers one by one, as they continue to elevate their performance.

> The third dimension is key to transforming organizational performance from good to great!

27.3 What Are These Layers?

At this time, CAMMP™ incorporates four advanced topics or layers. However, some organizations may decide to move some of these layers, such as sustainability, to the second dimension as an additional function. Organizations could also add more layers, such as logistics and supply chain, or anything that would be of value in their environment.

The third dimension topics are project success, competence, sustainability, and best practices. These are the advanced topics discussed briefly in the following chapters.

Please note, the coverage in the next chapters is brief and acts only as an introduction (with the exception of project success). However, for each topic, we will provide a reference to a leading professional association or organization that champions one of these subjects.

Chapter 28

Project Success

28.1 Introduction

Project success is vital, especially with the gaps in practice. This chapter will start by explaining the gaps, then go on to discuss the different perspectives related to success and explain the CAMMP™ approach. Section 28.5 will provide guidance on applying the concept.

28.2 Gaps in Practice

Some project management practitioners view project success from a narrow angle, such as from the perspective of a service provider or a technical project manager, but not from the viewpoint of the project *owner,* which is the reason the project exists in the first place.

Furthermore, lack of establishing project success criteria is another gap in common practice. During the authorization of the project, project and executive management should define and agree on the criteria to determine project success. It is common that some organizations ignore establishing success criteria, do not perform a proper definition of success, or do not communicate it to the project team.

The threat here is that, with no criteria, how can an organization assess project success and learn for the future? If organizations cannot assess, then how can they ensure proper project governance and a sound investment of the shareholders' funds?

28.3 Perspectives

28.3.1 Strategic and Organizational Perspective

Earlier in the book, we stated that CAMMP™ takes a strategic perspective for managing projects. For measuring project success, it is also necessary to take a strategic approach at the organizational level.

What does this mean?

It means that organizations should not limit their views to the *project* life cycle but must go beyond that, touching on the *product* life cycle. In other words, a project is a business venture, rather than an output that the project team produces before demobilizing. Ideally, project owner organizations must use an asset management process covering the whole *product* life cycle. However, for now, another way to think of this approach is that *CAMMP™ covers the product delivery*—the acquisition part of the product life cycle.

28.3.2 Project Owner Perspective

In line with the above, it is important to emphasize that, when it comes to success, CAMMP™ views projects from the project owner's perspective. It also takes the view of the organization, rather than the project manager or the project management function.

- For service providers, success might be limited to delivering the output within the client's requirements and realizing a profit.
- For the project management function in organizations, success might confine itself to delivering a good output that complies with the specifications and is delivered on time and within budget.
- However, for the project owner, success is more complicated than this, as explained next.

28.4 CAMMP™ Proposed Approach

In line with the above, CAMMP™ defines four dimensions (measures) of project success that the project management team and the organization can use. These four measures apply on the project level, but the first two can also be used for stage success.

28.4.1 Product Delivery Success: Technical

This first dimension of success assesses whether the team delivered the product (output) of the project successfully; this is about technical success. In other words, did the team:

- Complete the work?
- Deliver the product?
- Meet the technical standards and specifications?

Figure 28.1 presents this view, highlighting the following key points:

- The client (internal or external) works with the project management team to determine the success of the product delivery.
- The team can measure this dimension with a focus on quality, scope, and specifications.
- The measurement is in comparison to the project detailed plan (PDP).
- Because measurement is per the PDP, then this is mostly a technical assessment of the product compared to the specifications.

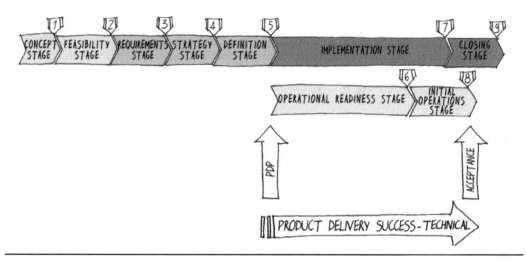

Figure 28.1 Project success, first dimension: product delivery success.

28.4.2 Project Management Success

Figure 28.2 shows the second dimension, project management success.

This dimension assesses whether the project management team delivered the project successfully per the PM plan.

- The focus here is primarily on time and cost, in addition to any other compliance requirements.
- Did the team deliver the project per the organizational project management standards?
- Did they deliver it per the established performance metrics and parameters?
- The client (internal or external) works with the project management team and project management office to determine this success.

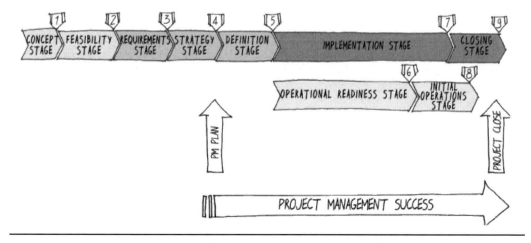

Figure 28.2 Project success, second dimension: project management success.

186 Project Management beyond Waterfall and Agile

- The team measures success in this dimension in comparison to the PMP.
- The team performs the success measurement during the Close Stage, with SG9.

28.4.3 Special Notices

There are two important things to note before continuing.

- **Note 1:** Notice that in discussing the second dimension, there was no mention of the project's being ahead of schedule or below budget. Rather, the terms used are "per the established metrics and parameters." Why is this? The answer would be a lengthy discussion, but let's just say, when management insists on below-budget and ahead-of-schedule performance, they are indirectly evoking the survival instinct of the project management team, which leads to their inflating the cost and schedule estimates. Inflating these estimates becomes a self-fulfilling prophecy, and the final cost is likely to end up higher than necessary but within the inflated budget. However, staying within the parameters, which could be ±5% or 10%, would allow for more realistic estimates.
- **Note 2:** The team can perform success measurement on Dimensions 1 and 2 before closing the project at SG9, and they would be able to include the results as a part of the project close-out report. However, Dimensions 3 and 4 would be a different story, as discussed below.

28.4.4 Project Delivery Success

Figure 28.3 represents the third dimension.

This dimension is the most difficult to explain. The purpose of the third dimension is to assess whether the team did *deliver the project successfully* per the various plans, specifically:

- Delivering product success per the project detailed plan (PDP) (the first dimension).
- Delivering the project per the PMP (the second dimension).
- Achieving other factors specified in the project authorization document (PAD).

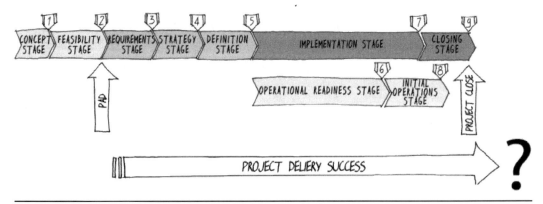

Figure 28.3 Project success, third dimension: project delivery success.

The first two bullets above refer to the first two project success dimensions, which are required input to this third dimension. Failing to deliver the product per the PDP and the PMP is a project delivery failure. To repeat: if one of the dimensions is not met, CAMMP™ considers project delivery is not a success.

Now, assuming the first two dimensions are in line with the criteria, the third bullet refers to other factors, which would be project or organization specific.

For example:

- In the case of an industrial plant, the success criteria can include customer satisfaction, achieving plant capacity within three months of final acceptance, no issues or gaps in the operating and maintenance procedures, etc.
- In an information system project, the additional criteria might be the end users' feedback and satisfaction rating.
- In a marketing project, the third dimension could include good feedback on the marketing campaign and customer satisfaction.

When can the team assess the project delivery success?

It is possible in some situations to evaluate this dimension at project close, but it is not likely. For most projects, the organization might have to wait months or even years to be certain about this dimension, which is represented in the "?" symbol in Figure 28.3.

The organization can measure the third dimension in comparison to the PAD.

Some practitioners might say this is not the project manager's responsibility.

We agree, especially because it might not be possible to assess success per this dimension before demobilizing the team. However, remember, CAMMP™ is a project management and delivery approach that takes the perspective of the *project owner's organization*. It is not limited to the effectiveness of the technical project management, nor it is a project manager performance assessment tool. Therefore, measuring project success is primarily an *organizational* activity rather than a *project manager* activity.

The person mainly responsible for this dimension would be the project sponsor, since that is the person who issues the PAD. However, it might also be the project management office or strategic planning, depending on the organizational preferences and how the OPMS is set up.

28.4.5 Objectives Success

The fourth and last dimension is the business[1] or objectives success, illustrated in Figure 28.4.

This dimension is designed to assess whether the project outcome resulted in delivering the expected capabilities, which in turn led to realizing the benefits that the organization outlined at the start of the project. In other words, this dimension has a direct link to the concept of *benefits realization*.

---

[1] The use of the term *business* here is in the general context and is not specific to for-profit projects. If the project is for a not-for-profit, non-governmental organization (NGO), or a government, then the driver for the project is a social or community need, not financial returns.

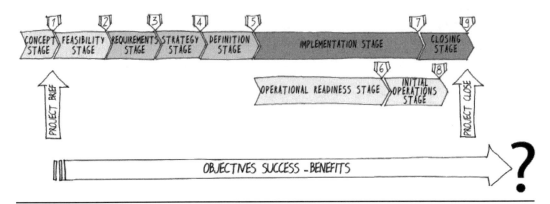

Figure 28.4 Project success, fourth dimension: objectives success.

Some may argue that this is the same dimension as project delivery success. On some projects, this may be the case, but not always.

Why?

The fourth dimension is concerned with the expected benefits and the justification behind the project. The third dimension, by contrast, is in relation to the PAD. There is a possibility that the PAD was wrong, and the sponsor did not develop the PAD properly to reflect the essence of the project. Therefore, the project team could deliver a good project according to the PAD, but from the organization's perspective, they delivered the wrong project.

If one refers to the examples that were used in the third dimension and reflect on them per the fourth dimension, we notice the differences.

- In the case of an industrial plant, the success criteria could be achieving the return on investment.
- In an information system project, the criteria could be reducing staff or increasing productivity.
- In a marketing project, project success could be increasing leads by x percent.

Another Perspective

Management may accept and push for an idea, an "executive pet project," and even intimidate the team, directly or indirectly, to take shortcuts in the feasibility study. Here again, they could end up successfully delivering the wrong project.

If the sponsors or executives have not erred in authorizing the project and the project as conceived is real and justified, failure to deliver the desired outcome can still occur somewhere, and the organization does not realize the benefits. The failure could be in the discovery, development, or delivery phases.

Finally, because of the nature of this dimension, it is unlikely that the team could measure it at project closure. The organization would only be able to measure objectives success after a lengthy period of time, which, in an extreme case, could be as late as the end of the *product* life cycle. An example of this type of occurrence follows.

28.5 Applying the Concept

28.5.1 Case Study

A client, an organization that sells valves and pipes for the petroleum industry, wants to build a gymnasium (gym) at one of its warehouses (facility).

What is the project?

Most will say "the gym."

Is it?

Is the company in the business of building gymnasiums?

No!

OK—then what is the project?

Maybe we should ask the question differently: What is the business objective of the project—*the driver*?

Oh! Now we can say: improve employee health.

Great; now it is evident. So how to measure success?

- **Product delivery success**. Did the team deliver the gym per the PDP? Meaning, did they deliver the building and equipment per given standards and requirements? If yes, good, the first measure achieved.
- **Project management success**. Did the team deliver the gym per the PMP, within the time and cost parameters? If yes, wonderful, another measure achieved.
- **Project delivery success**. Did the team deliver the gym according to the PAD? Another way one can consider this is, did the team deliver an acceptable and safe standard gymnasium? If yes, then excellent: The organization achieved the third measure. One can also include, did the staff like the gym and demonstrate this by using it?
- **Business objective success**. Did the project contribute to improving employee health? Proper measurement cannot take place until many months or even a few years after acceptance of the gym.

In the following, let's look at the same project from different angles.

28.5.2 Project Brief and Project Success

Remember, the project brief includes about three sentences: What is the project, why it is being pursued, and how does it align with the organization's strategy. In other words, it is brief.

Scenario 1

If you are a project manager, and *management mandates you to build a gymnasium,* would you be thinking of the fourth aspect of success, which is improving employee health? Or, would your focus be on building a gym project? Focusing on building the gym means you will be focusing on the facility only. What do you think?

Scenario 2

Now let us ask the same question with a slightly different idea statement.

Management mandates you to deliver a project to improve employee health (without mention of a gym); what will your project look like? It could include a gymnasium or not; it might include subsidized memberships to health clubs, and it is likely to include an awareness campaign about good health, or maybe sessions on healthy eating or cooking.

Scenario 3

In this third scenario, we will ask the same question differently.

In the project brief, *management mandate is to improve employee health by building a gym*. This project will likely include elements from the two earlier scenarios. The project manager will focus on building the gym and conducting a health awareness campaign. With this latter scenario, the chance of success—*business objective success*—is much higher.

As one can see from the above, changing the project brief slightly—even by a few words—can make a huge difference to the project.

Do you think that achieving the business objective for this project would be more likely with Scenario 3 versus Scenario 1 or 2?

28.6 When to Define the Criteria for Success

When does the team define the criteria for project success?

On the project level, which is the primary focus of this chapter, the team must define the success criteria as early as the feasibility study, but no later than project requirements.

In term of process, Chapter 17 on project management processes added a step in the authorize process for defining the success criteria for every stage and included another step in the close process to assess the success of the given stage formally.

If the team does not define the success criteria for the various dimensions, the project will fall into the trap of "completion = success," which is not effective from an organizational perspective and will not help the organization achieve a higher level of project management maturity.

28.7 Closing Comments

Let us emphasize a few points.

- CAMMP™ is a methodology for managing projects that is built on a project life cycle platform.
- The project life cycle starts with the idea and project brief and ends with Stage Gate 9, with project close, and after project acceptance by the client, internal or external.
- This project life cycle may include an initial operation period based on the type of project.
- This initial operation period is likely led by operation personnel, not the project team, but again, this is a function of the project domain.

This chapter covered the four dimensions of project success. The team can evaluate the first two dimensions during the project life cycle, before project closure. However, the organization cannot evaluate the last two dimensions until after SG9.

The project life cycle includes the three phases of a project—discovery, development, and delivery. This time span falls under the responsibility of the project sponsor and the project manager. However, as far as the organization is concerned, the *project* life cycle is only the acquisition phase of the *product* life cycle. In other words, the acquisition project is over, but the product life is continuing.

Who will handle the success measurements for the third and fourth dimensions, when the project sponsor and project manager have been reassigned? It could be the PMO, project management department, strategic business development, operations, or whatever department owns this process. The third and fourth dimensions are crucial to organizational success and proper governance, yet leading organizations continue to miss these dimensions!

Finally, why four measures? With each measure focusing on an area related to the project and the various teams that would be involved, it would be good to identify and pinpoint the failure, should one occur. This is crucial, but the aim is not to find a person or a team to blame, but to determine where the gap was. If a few projects are experiencing similar challenges with failures in one or two of the dimensions, that could be a system issue rather than a project or a people problem. Once the organization can identify the root cause(s) of failure, management can work toward building lasting and sustainable solutions and incorporating the new learning into the organizational project management system to avoid repeating in the future.

Chapter 29
Competence

29.1 Overview

Competence is about the demonstrated knowledge and experience of project personnel, sponsor, project manager, and team members, each in their area of work and role. Although competence can be part of normal professional development, it requires a special focus, which is necessary to elevate performance. It is about demonstrating the ability to do something well and repeat the process; it is about proving performance.

Many organizations that are implementing a methodology for project management stop their work at the project life cycle or the process groups. It would be of high value for them to consider incorporating a professional development program based on a competency development framework. The competence layer is an advanced topic in CAMMP™.

29.2 Resources

Many organizations and professional associations have resources or guides on competence, but the leading references here are the IPMA® Individual Competence Baseline (ICB®) and, on the organizational level, the Organizational Competence Baseline (OCB®). Furthermore, many of the IPMA organizations around the world follow the ICB or have a modified version specific to their members.

GAPPS (Global Alliance for Projects Performance Standards) is another independent organization that offers standards that are role specific and build on competence. Some of these standards are for program manager, project manager, and project sponsor.

The same can be said for other associations that focus on a given area, such as value management, cost engineering, and project control.

29.3 IPMA ICB

Per IPMA®'s Individual Competence Baselines, there are three general categories of competence to consider for project teams.

- **People competences.** "These consist of the personal and interpersonal competences required to successfully participate in or lead a project, programme or portfolio."
- **Practice competences.** "These are the specific methods, tools, and techniques used in projects, programmes or portfolios to realise their success."
- **Perspective competences:** "Under this heading come the methods, tools and techniques through which individuals interact with the environment, as well as the rationale that leads people, organisations and societies to start and support projects, programmes, and portfolios."

This book will not expand on the details of the various elements for two reasons: (1) the depth of content in ICB and these other resources; (2) the fact that competence is highly dependent on the various roles within project management.

29.4 Common Misunderstanding

One can often observe a common misunderstanding in the professional community in confusing *competence* with *knowledge*. For example, the claim that some certifications are competence based is questionable and debatable. The simplified explanation is that knowledge can be tested in a multiple-choice exam, whereas to assess competence one has to show evidence of know-how rather than memory-based responses. Proper assessment of competence is done by qualified assessors, on a one-on-one basis. This one-on-one

Case Study

The author had an opportunity recently to deliver a workshop to a group of project engineers in the oil and gas industry. During the workshop, they shared with me information about competency assessments they all had to do.

Their management hired a consultant to perform the assessment and determine their level of competence, and the result would be used for their development toward project manager positions. They were furious. Here are some of the issues:

- The assessment interview was only 20 minutes per individual.
- He gave them a score of 1 on every category, which seems like a copy–paste job.
- In the assessor's view, they only needed to achieve a 2 or 4 (on a scale of 1 to 5) in all categories.
- The categories were limited solely to the *PMBOK® Guide* processes.
- There was no assessment specific to their other areas of work, such as engineering, construction, FEED development, etc.
- There was no assessment about HSE or related topics.
- There was no assessment in comparison to their internal processes, although the consultant was given that information to use.
- Finally, to develop them, he was recommending courses from his own organization, including PMP® boot camp training.

Unfortunately, this situation reveals the consultant's lack of competence, or should we just say, his incompetence? Also it is unethical, in our humble opinion, because he ignored the company's internal processes and only used a generic approach. The sore point for the project engineers was the reference to the consultant's internal courses only, with no mention of other development activities, on-the-job training, and coaching or mentoring.

assessment is the approach that IPMA uses to grant the higher-level certifications, through formal assessment and usually by more than one assessor. Candidates for an IPMA certification have to prove their abilities.

29.5 Competence in Organizations

29.5.1 The Current State of Practice

These are some of the current "state of practice" factors related to competence:

- Many organizations today depend on general project management training, which is often generic and not tailored to their environment.
- Some do not even have a formal professional development program and depend on their staff's getting training on their own or pursuing generic certifications that might not align with the organizational system, assuming one exists.
- Another gap in practice today is that many organizations do not even have a formal organizational project management system (OPMS) or a methodological approach to use as the basis of a competence baseline.
- The human resource system in many organizations might not even recognize project management–related positions. Consequently, there is no basis for formal learning and definitely no competence development framework to use for professional development and performance assessment.
- Some organizations do not fully understand the difference between knowledge and competence.
- Some might have the basics of an OPMS, and might even have some focused training, but not at the level of a competence-based framework.

The above may vary from one country to another and one organization to another. However, it holds true for a large number of organizations, especially those who have embraced project management in the last decade or two.

29.5.2 Implementing Competence Development

It is critical not to take the topic of competence for granted or equate it to general professional development. Many organizations, including major global powers, have not yet adopted a competence-based development framework for elevating their staff's performance, at least in the project management domain.

Once an organization respects project management and regards it as an independent domain, it might go through stages from ad-hoc training to organized training, then formal development and recognition in the human resource system. The next level would be to develop a professional development and leadership program that is based on a competence development framework, and even certifications based on competence. As mentioned before, IPMA is a leader in the general project management category, but many other organizations focus on specialties within the field of project management. These include AACE, SAVE International®, Guild of Project Control, and many others.

29.6 Competence and CAMMP™

The author has been fortunate to work for a leading organization in which he had the opportunity to be part of the competence development framework for various positions in the organization's project management functions, such as project engineer competence elements and cost engineer competence elements.

What would be required here is for an organization to establish a competence development framework in relation to the CAMMP™ Model. Professionals need to understand:

- The various concepts presented here in terms of differentiating project life cycle from process groups and all other elements.
- How to tailor a project life cycle model to a specific project environment and project type.
- How and when to apply the processes and their various steps.
- How to establish project classification criteria to determine what life cycle model would be the proper approach for a given project class.
- How to develop the various stage and process deliverables and distinguish among them.

Numerous competence elements could be listed here, but in order not to re-create the wheel, IPMA and GAPPS would be great starting points for further reading and consideration. Their standards are available free of charge from their corresponding websites.

Chapter 30

Sustainability

30.1 Special Message

Dr. Joel Carboni, Founder of Green Project Management (GPM®) Global and President of IPMA® USA, provided some of the content of this chapter. Dr. Carboni also edited the chapter and other sections on sustainability at the request of the author. Dr. Carboni is a leading advocate of sustainable practice in delivering projects.

30.2 Overview

The world in which we do business is changing; disruptive business models, globalization, climate change, and the rise of millennials are impacting the global market and forcing changes that require greater adherence to sustainable practices. At our current pace, humanity is consuming 1.6 percent of the planet's renewable resources annually, and the negative impact on ecosystems and societies is becoming more severe. Therefore, it is vital that we consider sustainability in everything we do.

Organizations drive change via projects. These projects could be related to facilities and impact people and the planet, or they could be internal projects touching on people and prosperity. Project management is necessary to manage these projects.

Consequently, cannot project management skills be used effectively to manage change and deliver projects with a modern mindset, with sustainability as a core principle?

30.3 Resources

GPM Global was established in 2009 with a mandate to focus on sustainability principles and deliver projects with sustainable approaches. GPM has built on the concept of the triple bottom line of people-planet-prosperity and added to this idea a focus on product and process. This emphasis led to the P5™ Standard: The GPM Global *P5™ Standard for Sustainability in Project Management* (GPM Global 2016). GPM released an update to the first edition in September 2016.

In addition to the P5 Standard, GPM also released PRiSM™, Projects integrating Sustainable Methods, and the associated *GPM Reference Guide to Sustainability in Project Management*, published in 2013 (GPM Global 2013).

P5 and PRiSM are the foundations for the GPM certifications, along with various ISO standards and UN initiatives.

30.4 P5

The P5 Standard includes the five core elements of society (people), environment (planet), financial (prosperity), maturity and efficiency (process), and objectives and efforts (product) (GPM Global 2016). The standard further breaks down each of these elements into sub-elements, such as child labor, cultural impact, organizational learning, transport, water consumption, waste, increased business flexibility, cost–benefit ratio, and local economic impact. It also provides recommendations both on practices that project professionals should consider and on the benefits of taking specific actions.

Please refer to the P5 Standard for further reading.

30.5 PRiSM™

PRiSM and its associated guide are published as a stand-alone method, which uses ISO 21500 process groups as its project life cycle. The method includes various deliverables focusing on sustainability that the project team should address as they work on the project. These deliverables include a sustainability management plan. PRiSM is an excellent tool and provides the basic elements of managing projects with the right mindset of sustainable change. The only concern with PRiSM is that it uses the ISO process groups as the project life cycle, and earlier in this book, it was clearly stated that the process groups are not project phases. It is our understanding that an update to PRiSM, in progress at the time of writing, will fix this gap.

Consequently, the author's view is that sustainability is a core function more than a stand-alone project management method. It would be best to link PRiSM to a method like PRINCE2® or CAMMP™. Therefore, the author incorporates sustainability as a core function and layer in the third dimension of CAMMP™.

30.6 Sustainability and CAMMP™

If one refers to Part E, one will notice that the feasibility study of CAMMP™ incorporates sustainability considerations, which is equal in importance to the financial, cost, schedule, and market considerations. Instead of limiting the work to conducting an environmental impact assessment, organizations must move toward a sustainability impact assessment.

As the project proceeds along the project life cycle, all of the other CAMMP™ stages include a section on sustainability. Defining the product and project requirements must include the sustainability requirements, the project management plan must include a sustainability management plan, and the project detailed plan must define all sustainability elements. Similarly, the same concepts apply to all other work, in all phases and stages. What is described here is the approach used in CAMMP™ to integrate the third-dimension topics into the first two dimensions and embed them in the various deliverables at the project and stage level.

Chapter 31

Best Practices

31.1 Overview

As was highlighted earlier, in Chapter 3, the *PMBOK® Guide* states that it is about *good* practices, not *best* practices. Does this mean best practices, or leading practices, do not exist?

In the *PMBOK® Guide*, where the term *best practices* is used, it refers to domain-specific practices, because it not logical that there would be generic best practices. Therefore, guides such as ISO 21500 and the *PMBOK® Guide* provide the good practices of project management in a generic format. These good practices provide the foundation for organizations to build on and work toward implementing best practices. The author is not a fan of this term and prefers *leading practices* over *best practices,* since what is best in one context might not be best in another. However, we use the term here because our main reference is CII, and CII uses best practices.

Therefore, it is important to stress that best practices are unique to specific industries. For example, in some sectors, such as capital projects,[1] there is a set of best practices that can be applied to the project life cycle and, possibly, to every stage. These include practices such as constructability, change management, zero incident, and other topics.

31.2 Resources

A good reference for capital projects is the Construction Industry Institute (CII). Other industries will most likely publish their own best practices guidelines. At the time of writing, the CII website (construction-institute.org) lists 17 best practices for the capital projects sector. Some of these are:

---

[1] Capital projects are capital investments projects, such as facilities, which require engineering and construction.

- **Planning for startup.** This would be related to Operational Readiness and Initial Operations in CAMMP™.
- **Planning for modularization.** This topic relates to the story shared earlier about the mega project on an island (page 71). (At the time of that project—the mid-1990s—this topic was not a recognized best practice.)
- **Front-end planning.** This best practice is related to the various efforts that the team would perform during the CAMMP™ Development Phase. In the capital project industry, the project detailed plan would carry the title "front-end engineering design."
- **Change management.** This is also a best practice. In this industry, changes could be the number one cause of project failure; hence the need for particular attention to change offered by the CAMMP™ Model.

The above apply to one given industry; further details would deserve their own book, and readers from other industries might not be interested. However, it is vital to stress that each of the above topics has in-depth coverage published by CII. All these resources use research from real projects submitted by CII members' organizations.

31.3 The Value of Best Practices

The following quotation is from the CII website: "A CII Best Practice is a process or method that, when executed effectively, leads to enhanced project performance. CII Best Practices have been proven through extensive industry use and/or validation."

CII claims that the effective application of every best practice can reduce the total project cost 1–3 percent per best practice, which would translate to millions of US dollars on large and complex projects.

31.4 Best Practices and CAMMP™

Best practices, if they exist for a given project type, would be applied across the project life cycle. However, most best practices are applied early in the Development Phase, while the project is still in the early stages and before reaching the Delivery Phase. Most of them could be in the Requirements Stage, because that stage sets all of the requirements that would be used both in the upcoming stages and for the project estimates of cost and time. Some best practices could be applied initially in the Requirement Stage and updated in the Definition Stage.

Now, of course, elements of these best practices are implemented throughout the project. For example, the change management best practice, the policy, and the guide would be defined early on, even as early as project authorization. However, such practices would be complied with every time a change is under consideration. The same concept applies to most of the other best practices.

Consequently, best practice is a layer in the third dimension of CAMMP™ that consists of various steps applied along the project life cycle.

Section III, Part G: Practical Management of the Three Dimensions

Chapter 32

Managing Across the Stages

32.1 Introduction to Section III

How to manage projects efficiently?

How to deliver products successfully?

How to reach best-in-class performance?

The message of this book is that individuals and organizations must go through a paradigm shift to recognize that the journey to excellence is not an easy one and cannot be achieved with over-dependence on common sense and accidental project managers.

Some organizations manage projects per the second dimension only, following the *PMBOK® Guide* or ISO 21500 process groups. Others manage per a project life cycle or a software development life cycle, which is good, but not enough except in special situations. The third dimension adds value and helps to reach a higher level of maturity. *Once again, the journey to excellence requires some form of integration of all of these concepts.* Consequently, how can one manage per the three dimensions, while keeping things as simple as possible?

In response to a recent post, a professional colleague answered this question with the following: "A three-dimensional model will not work with one-dimensional thinkers." Maybe he was right. However, would we agree that "one-dimensional thinkers" will have to settle for less than optimal outcomes?

Consequently, how can one manage per the three dimensions, while keeping things as simple as possible?

Again, it might be acceptable to organizations new to project management to use one of the dimensions without the others, but sooner or later they have to consider elevating performance.

On the other hand, organizations that are already comfortable with project management can consider the approach presented here as an opportunity for elevating performance, today!

In Section II, the focus was on presenting the concepts and explaining the various dimensions and their components. In this section, we shift toward topics that will help the reader visualize the application of the three dimensions in the real world.

32.2 Reflections on Challenges

Some practitioners of project management raise their concerns in dealing with the challenges they face in project management, especially when relying on only one guide or resource. The following comments reflect some of these difficulties.

- "I understand the guide, but now I need to apply this to my specific project (or industry)."
- "How do I connect the dots?"
- "I know the process groups repeat, but how do I implement them?"

> The project management processes are better viewed for managing a phase or stage, not the project.

In general, these challenges arise because many practitioners do not understand that they have to think at two levels: managing the *stage* level and managing the *project* level—these are the first two dimensions of CAMMP™. They also need to think of elevating performance through the third dimension.

Keep in mind that the ISO and PMI guides present the processes to manage a project or a phase. However, as explained in Chapter 17, *the processes are better viewed for managing a phase or stage; not the project.* The use of the process groups is not enough for projects that are not simple or small. Therefore, a project manager and the project management team can manage the phases and stages using the process groups, but to manage "across the stages," they must depend on a project life cycle model.

Let's maintain the focus on the first two dimensions now. *How do we manage across the stages? What does it mean to "think at two levels"?*

Remember what was presented in earlier chapters—there should be a stage charter (stage authorization document, SAD) and a project charter (project authorization document, PAD). Similarly, there is a stage management plan (SMP) and a project management plan (PMP); a stage detailed plan (SDP) and a project detailed plan (PDP); and so on.

For example, control must occur across the project life cycle, as the team progresses from one stage to another; yet control is also within the stage as the team moves between the processes. Similarly, risk analysis is for the project as a whole and for each stage, because there are *project* risks and *stage-specific* risks. All the chapters in this part will help the reader visualize how to think at two levels for some of the functions. However, before the dedicated chapters, it is necessary to present a couple of points to help visualize the integration of the first two dimensions, the project life cycle, and processes.

32.3 The Charter versus Managing Across the Stages

The author once posted a question on LinkedIn, asking, "How many charters are there on a project?" A blog article addressed this topic as well. Notice the trick in the question; the question was not about how many *project charters,* but how many *charters along the project life cycle.*

This question generated active discussions, and most of the responses did not align. Most said one charter; a few said one charter per project updated with every phase; and a few agreed with the position of CAMMP™.

The proper answer is that there is one *project* charter, but, if one understands ISO 21500 and the *PMBOK® Guide* correctly, a *charter* is about authorizing the project or phase. That means every phase must have an authorization, regardless of whether it is called a charter.

To elaborate further; the project charter, per the guides, is issued *after* management makes the decision to proceed with the project. This decision is typically made after a feasibility study is performed. The feasibility work is pre-project and pre-charter, per these guides. However, would the reader agree that the feasibility study is carried out as the result of management's approving the idea of the project and the business case?

The above means that the project feasibility work should be a stage of the project owner's project life cycle. As addressed earlier, the project feasibility is a stage in CAMMP™. In this regard, *can a team start working on the Project or Feasibility Stage without management authorization?* Is this authorization for the Feasibility Stage not a stage charter? This stage authorization (charter) is published weeks or months before the project charter. Remember, the key question is: does this authorization (of the feasibility study) not reflect the intent of the Develop Charter process? Figure 32.1 presents this concept for the need of a project charter and stage charters.

Let us summarize using this example:

Strategic business planning in a healthcare organization has an idea—a business case—to build a new hospital in the recently developed part of the city. They submit the idea to management; if management approves the idea, management must issue an authorization (stage charter) to a project sponsor or manager (lead) to start the stage work and study the feasibility of such a project. After authorizing the stage (Authorize process), the team will plan for

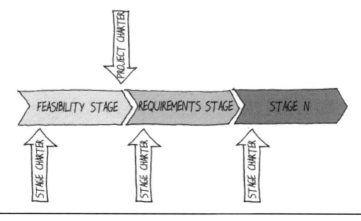

Figure 32.1 The project charter and stage charters.

conducting the study (plan scope, time, cost, risk, procurement). Once the plan is approved, the team will conduct the feasibility study (implementing the stage work). They will also control as they move along and close when done. This is an example of applying the main processes within a stage that the guides consider pre-project.

If the project is feasible and management decides to proceed, they will issue (through the sponsor) a project authorization document (project charter). This charter will clearly state their intent to move forward with the project until completion. However, management is only authorizing the next stage, and they can do so via a specific stage authorization or use the project authorization. The project management team (PMT) would plan and proceed to execute this stage work, which is developing the project requirements (processes repeating). When the Requirements Stage work is complete and approved, the team will repeat the project management processes for every upcoming stage.

In closing this example, the project charter must be fixed once it is issued, and it does not need to change unless the objective of the project changes. However, the organization still must authorize every stage of the project. This stage authorization practice is necessary for proper governance.

32.4 Simulating Managing Across the Stages

32.4.1 Overview

Figures 32.2, 32.3, and 32.4 are from another book (in progress), which is a sample (a case study) for a large and complex project. The presented figures show the Discovery Phase, Development Phase, and Delivery Phase, respectively.

In Figure 32.2, the focus is on the Discovery Phase—in particular, the Feasibility Stage. The stage charter, in this case, is the stage authorization document (SAD) before PG1 (Process Gate 1). The project charter is an output of this phase and is the project authorization document (PAD) shown on the right side before closing the Feasibility Stage and moving onward to the next stage, which would be the Requirements Stage, shown in Figure 32.3 (page 210).

32.4.2 General Explanation

These three figures are per a project life cycle slightly modified from the standard model. In this case, there is a stage to develop the master plan (between requirements and management planning), because this project is to develop a large facility. Because of this addition, there is one more stage gate—the Master Plan Stage, which is in the Development Phase presented in Figure 32.3.

In Figures 32.3 and 32.4, every "column" represents a stage, and it should be obvious how the processes repeat.

- The top part is for authorizing each stage with an SAD and a process gate, PG1.
- The next block represents the stage management plan (SMP) and its process gate, PG2.
- After the SMP, it is time for the stage detailed plan (SDP) with PG3.

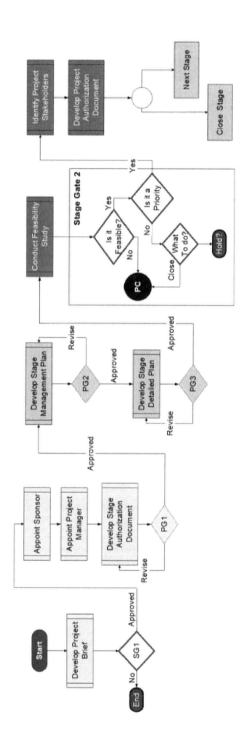

Figure 32.2 Sample large-complex project, Discovery Phase.

210 Project Management beyond Waterfall and Agile

- After the management and detailed planning processes, it would be time for the implement process. Implementing the work varies in relation to the scope of the stage. The process in the image shows the deliverable that would be produced for the given stage. These are the feasibility study, project requirements, master plan, project management plan, and project detailed plan (which is called *preliminary engineering package* in this case study).

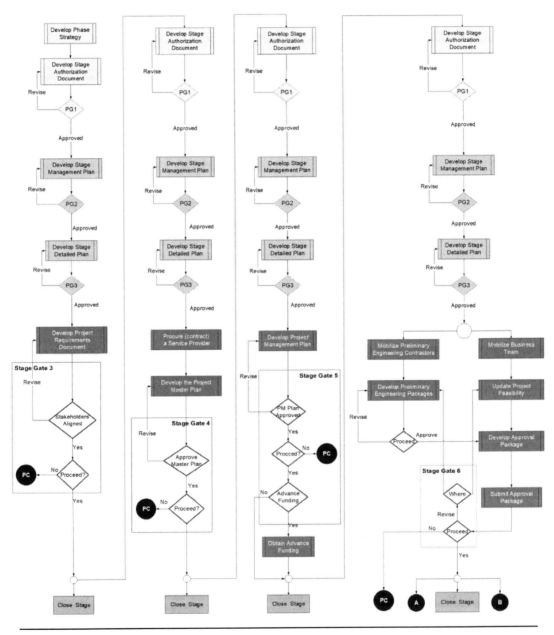

Figure 32.3 Sample large-complex project, Development Phase.

Managing Across the Stages 211

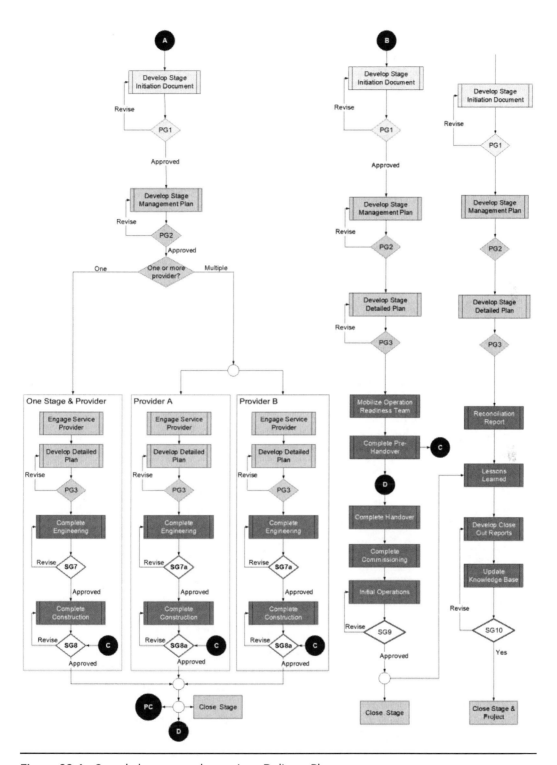

Figure 32.4 Sample large-complex project, Delivery Phase.

- Notice that each stage gate is actually more than one step but clearly shows the possibility of stopping the project at any of these stage gates.
- At the bottom of each column are the close stage processes, but only after the stage gate for the given stage.

One can also notice that in each of these columns, at the stage gates, if the project is stopped, there is a circle with "PC," referring to *project close*. All of those circles are references to start the Close Stage, the right-side column in Figure 32.4.

The following comments are unique to each image, starting with Figure 32.3.

- The top left column shows the Develop Phase Strategy. This step is added here because for large-complex projects—and, this being a critical phase—it would be necessary to have part of the project management plan—the strategy section—considered at this time, but only for the upcoming phase.
- The last stage in this phase, on the right of the image, assumes that the preliminary engineering work will be outsourced. As a result, the stage detailed planning is part of the service provider scope.
- Also note for this stage the mobilization of the business team to work in parallel with the preliminary engineering team to update the project feasibility and commercial model, along with preparing the project funding request.

In Figure 32.4, for the Implementation Stage, there are two scenarios presented. One shows one service provider performing all of the implementation work, and the other shows two providers (could be more). In either case, the assumption is that one or two providers will perform the full EPC (engineering, procurement, construction) work.

Also, in this case, there is no pilot or Initial Operation Stage as an independent stage; it is shown with the Operational Readiness Stage.

32.4.3 Summary of the Project

The above represents a large-complex project. What does this classification mean? Refer to Chapter 41 for discussions on project categorization and classifications. However, what is important to stress here is something that has been stated more than once in this book. *Do not let the number of processes and repetition of the processes intimidate you.* For once, on these large-complex projects, you are not alone—there should be a team consisting of various project management personnel and technical subject matter experts. It is not uncommon to have a project management team comprising tens of professionals and project teams composed of hundreds of staff members. Construction of large-complex facilities may even require thousands

> **What is a Large-Complex Project?**
>
> It is hard to define a large-complex project without the context of the performing organization and the type of project. If we use a generic definition, a large-complex project for one organization can be a small project for another.
>
> In general, a large-complex project would cost millions or hundreds of millions US$, take more than one year to complete, and require numerous resources and team members (hundreds or more).

of workers. Still, project management processes would be similar. Just treat every stage as a sub-project.

Furthermore, with the proper investment into an organizational project management system (OPMS), many of the common steps repeat, and the team could depend on CPM (copy-paste-modify) of the generic information that is highly dependent on the OPMS. What is unique and changes from stage to stage and project to project is the product-oriented processes that would require significant effort. The next example is about building an organizational project management system.

32.5 A Project Example, End-to-End

32.5.1 The Project

Let us assume that the organization is the ministry of foreign affairs for a given government. The project is about building a sustainable OPMS. This OPMS will be implemented in the home country across the various divisions and globally for any projects led by the various embassies, such as aid projects.

32.5.2 Discovery Phase

As the new minister takes charge after a government reshuffle, she recognizes that the ministry has numerous projects all over the world, but the reports on these projects are entirely different. The reports do not have consistency in terms of format and content. Furthermore, it is not clear where these projects stand or how they are performing. After further investigation, she finds that the ministry does not have a standardized process for managing projects and programs. Because the minister comes from a project management background, she asks one of the ministry's senior staff to draft the project brief for building a comprehensive and sustainable OPMS, as was described earlier. She approves the project brief and appoints a project sponsor to lead the effort.

The sponsor implements the Mobilize Project Manager process (Chapter 17), which leads to hiring an expert project manager who is certified in the CAMMP™ Model. The sponsor authorizes the project manager to lead the project, but starting with the feasibility study. The sponsor also sets the success criteria for the stage.

Because the project manager has the authorization (a Feasibility Stage charter), the next step is to mobilize a project management team to help him develop the stage management plan. At this stage, a couple of people could be enough—a senior planner and a deputy project manager with experience in the ministry type of work. To develop the stage management plan, the project manager follows the Plan Management process described in Chapter 17, which includes all of the process steps necessary to manage the stage work.

Once the stage management plan is complete, the project manager mobilizes a team of lead experts from the various divisions within the ministry; these include senior professionals from strategic planning, planning and budget, facilities, international affairs, information technology, human resources, sustainability, health and safety, etc. This team, with senior professionals, develops the stage detailed plan (SDP)—for the feasibility study only. The SDP includes all of the Plan Details process of Chapter 17, such as the comprehensive scope, cost and schedule estimates for the Feasibility Stage, and the resource requirements; defines the applicable quality standards; identifies and analyzes the stage risks; and develops response strategies.

Next, it is time for the project team to conduct the feasibility study, which would consider the various factors addressed in Chapter 19. The team also will follow the Implement process from Chapter 17.

Throughout this study, the project management team controls the stage work to ensure that the team does not lose focus and is not doing more than is required or less than is needed. (The Control process was also defined in Chapter 17, page 103.)

Once the feasibility study is complete, it is time for a stage gate and the sponsor's approval.

After consultation with the minister, the project is approved, and the sponsor issues the project authorization document (PAD). The minister's approval is the first approval point (refer to Chapter 33). With this approval, it is time to close the stage with the aid of the Chapter 17 Close process.

32.5.3 Reflections

It would be valuable to reflect for a moment.

Except for the feasibility study itself, all of the work, such as the stage management plan and stage detailed plan, were particular to the stage: estimates for the stage, risk for the stage, procurement for the stage, quality for the stage. This approach represents thinking (and working) at the stage level.

Let's shift focus to thinking on the project level. The feasibility study was about the project. Part of this study were cost and schedule estimates; these were for the project, Class 1 Estimate, as will be discussed in Chapter 34. The study also includes the sustainability considerations, risk, and other factors. All of these actions focus on the project. All of the feasibility considerations will lead to the final recommendation for management decision.

32.5.4 Development Phase – Requirements Stage

The PAD approved this stage, and unless something is out of the norm, that authorization is enough to start the stage work.

The next step would be for the project manager to develop the stage management plan (SMP). It might be necessary to add staff to the project management team. As before, this plan is unique to the *stage only*. It includes all of the relevant steps.

From the SMP the team will continue work to develop the stage detailed plan (SDP). As above, the SDP is focused on the Requirements Stage, and it is based on all of the relevant steps.

With the SDP completed, the project manager will mobilize the necessary team members to conduct (implement) the stage's main scope, which is to develop the project requirements document (PRD) per the Implement process. The PRD will be prepared per the guidelines presented in Chapter 20.

Obviously, control is throughout the stage and stage closure once all is acceptable.

Once again, the main processes (process groups) and their applicable process steps would be specific to the stage, whereas the requirements document's focus is on the project. At the stage gate, management would be concerned with the PRD and that it complies with the project authorization; this is a control function on the project level.

32.5.5 Summary of the Rest of the Stages

Strategy Stage

After the requirements are defined and approved, the project management team will move onto the Strategy Stage and develop the project management plan (PMP), including all aspects and components, as discussed in Chapter 21.

Definition Stage

Next, it is time for the Definition Stage and the PDP, per Chapter 22. The PDP will include the scope of work for the Delivery Phase, which should include a list of all the project deliverables that the team will produce in the Implementation Stage. These deliverables would include all of the policies, procedures, guidelines, methods, and process packages that will have to be developed to form the complete OPMS. The plan will also include the quality standards that the team will have to follow, identify the risks and manage them, decide if they can do the work internally or through outsourcing, develop the cost estimate for the full project along with the schedule, and all other components of the PDP.

Implementation Stage

Implementation is easy, process-wise, but it is time consuming and requires significant effort. It is time to implement all of the work specified in the Definition Stage, such as writing procedures, developing process flowcharts, etc.

Operational Readiness Stage

In parallel to the implementation work, another team should be working on marketing and communication (MarCom) to prepare the organization for the upcoming change. This team will also be working on the roll-out plan, along with selecting champions and other personnel to support the roll-out in the various ministry divisions around the world. This work is part of the Operational Readiness Stage, discussed in Chapter 24.

Initial Operations Stage

Such a project can benefit greatly from a pilot in one of the divisions, ideally in the home country of the ministry. Any learning from the pilot could be used to modify some of the deliverables—deleting some and maybe adding others. Then, it would be time to roll out. The roll-out plan would have been defined earlier.

Close Stage

Remember, the team should have been closing every stage, and now it is time to close the project, summarize all of the lessons learned, and update the organizational records.

32.5.6 General Comments

Per the previous text, the project manager will continue to lead the project to completion and delivering success. Note the following:

- Along the project life, the project manager could have been changed, maybe more than once. Although this is not preferred, it might be necessary.
- The project management team will consist of a few members early on, but later it could grow, reaching a peak during the Definition and Implementation Stages. This team will include planners and schedulers, cost specialists, procurement personnel, among others.
- The technical team is actually multiple teams. The leads may be the same throughout the project, but the staff conducting the feasibility study are likely different than those developing the definition package or implementing the work.

32.6 Closing Comments

In the earlier example, authorizations on projects are at two levels—*project* authorization and *stage* authorization. The same goes for all other topics, from estimating and control, to change and risk management, to project success. Project teams (and organizations) must understand that *managing projects happens at two levels: the stage level, and across the stages* (managing the whole project).

Consequently, the two-level management means there are stage management plans (with an "s"), and there is *one* project management plan. There are stage charters and *one* project charter. There are stage estimates, and there are project estimates.

The next few chapters will further visualize the various concepts.

Chapter 33

Project Approvals

33.1 Introduction

It is quite common to meet professionals working on small projects or projects within organizations new to project management. These professionals often think that there is only one point of approvals for projects—one estimate, one plan—stemming from the usual confusion of thinking that *process groups* are *phases*.

It is likely that some organizations do have one only approval point, which might be enough for the types of projects that they handle. However, this is not a good practice for most projects.

Following are the various types of approvals required on projects.

33.2 Types of Approvals

33.2.1 Stage Gate Approvals

Stage gates are one of the approval points used on projects. What are these gates approving?

1. SG1 is about the project's alignment with strategic objectives and management's decision on whether to proceed with a feasibility study or stop the project.
2. SG2 is about reviewing the results of the feasibility study, and, if the project is viable *and* a priority, then authorizing the project.
3. SG3 is about ensuring the alignment of the various stakeholders to the project's requirements, and once there is agreement, the team will proceed to project management planning.
4. SG4 is for executive management to review the project management plan (PMP). If they agree with the approach, they will authorize the project to the next stage. This gate has another significant role, which is advance funding, if required.

5. SG5 is "the point of no return." Once the stakeholders agree to the project detailed plan (PDP), they are likely to grant the final approval along with committing the necessary funds to complete the project. Because full funds are committed, it is unlikely that the organization will cancel the project after this point—hence, the point of no return.
6. SG6 is to verify that implementation has progressed enough for the operation team to accept custody and control of the project's product from the project management team, which is the handover gate.
7. SG7 indicates completion of all of the implementation work, including any exceptions or snag list items. At this stage gate, the project manager will release the remaining members of the implementation team.
8. SG8 is the project's final acceptance from the client, which takes place after handover and completion of all the snag list items. The timing of acceptance is typically dependent on the project's industry (domain), and it is usually at the end of the initial operations stage, if such a stage was required.
9. SG9 is the final gate, reflecting that all deliverables up to and including the closeout report are complete, and the project is officially closed.

33.2.2 Process Gates Approvals

Also remember that Chapter 16 introduced the process gates—four process gates that repeat in every stage:

1. PG1 is the exit point for the Authorize process and indicates approval of the stage authorization document (SAD), which allows the project manager to start management planning for the stage.
2. PG2 is for the approval of the stage management plan (SMP).
3. PG3 is for the approval of the stage detailed plan (SDP).
4. PG4 is for validating that all the stage work is complete per the plan before closing the stage. PG4 can often overlap with the stage gate for that given stage, or, preferably, it would be the final point before submitting the stage deliverable to the stage gate review.

33.2.3 Budget Approvals

In a stage gate process, as in CAMMP™, it is necessary to have several approval points along the project life cycle instead of just one; these are the stage gates along the project life cycle, as discussed, and the process gates required within a stage.

It is worth noting that each of these points—process or stage gates—is approving something very specific for the project, such as producing a deliverable or exiting a stage and entering another. However, from a financial perspective, most of these gates have limited or no role regarding *funding the project*. Therefore, in the context of this chapter, the term *budget approvals* refers to another type of approval: typically, formal approval of funds.

Some might argue that having multiple approval points is bureaucratic; that may be so, but we cannot always control this. An earlier chapter addressed the question of bureaucracy, and we will briefly restate the explanation here. If an organization is highly bureaucratic, then it is only natural that project management will be as well, even if there is only one gate. In these situations, the issue of bureaucracy is an inherent problem that project management cannot fix. In a mature organization, passing through the various gates must be efficient, increasing the chance of project success.

> Bureaucracy is a function of organizational culture,
> not the number of gates.

33.2.4 Timing of Approvals

OK, nine gates, and we are still not clear on when the money is approved; when do we get approval for the project budget?

Here are some scenarios:

- In some organizations, the question of budget may not be clear or explicitly stated. The organization may have a dysfunctional project prioritization or approval process. Alternatively, it may treat projects as part of day-to-day operations without allocating a specific budget to a given project.
- In other organizations, management may keep the project budget confidential, not sharing it with the performing team.
- Some other organizations do not budget for their internal resources and budget only for contracted work and external resources.
- In other situations, executive management may have a predefined and constrained budget. In this scenario, the team is required to deliver the project within that set budget constraint.

However, if one shifts the focus to organizations that have a more structured, formal or semi-formal project management system, there will be an official funds approval process. Furthermore, final project (funding) approval is at one of the nine stage gates.

Which one?

Some organizational practices place the approval point somewhere between Stage Gate 2 and Stage Gate 5.

Is this the best practice?

Maybe it is for small and internal projects, although any approval of budget before Stage Gate 4 in the CAMMP™ Model is premature.

How about if the organization has a four-gate model?

Good point; Stage Gate 4 in the standard model comes between the end of project management planning and the beginning of detailed planning. Therefore, if the organization's project life cycle is only a four-gate process, funding approval would occur at the gate that would be equivalent to Stage Gate 4 in the standard CAMMP™ Model.

What is vital to stress here is the work leading to the stage gate, and not its number. In other words, what should the team accomplish before seeking and obtaining final approval?

33.3 Budgeting Common Practices

33.3.1 Preset Budget

One common practice is to establish a preset budget at the time of project authorization.

Some organizations, particularly governments, follow this approach for internal projects. However, there are risks with such an approach, because the figures (both cost and time) that the team establishes at this early stage of the project are not highly accurate due to the low level of project scope definition. As a result, the risk of missing the time and cost targets is high.

33.3.2 Project Budget (Internal and External Costs)

Another common practice is for the organization not to budget the early Project Stages, treating these as part of their overhead (operating expenses), and only budget for the external costs and contracts. The threat with this approach is that the organization will have only a high-level idea about the project budget until the work progresses enough to seek quotes or proposals. If the price is much higher than expected, management could cancel the project, and the effort that went into it is not recoverable.

33.3.3 Constrained Budget

One more common practice is for management to establish a constraint on the time or budget. If time or cost is a constraint, then the team has a different challenge, and in this case, the project scope or quality will likely have to be adjusted to conform to the preset constraint.

Although all of the above practices are not ideal, they are common practices and would be per organizational cultures or procurement policies. They might be acceptable for certain conditions.

33.4 Leading Practices

33.4.1 Is There a Leading Practice?

In this context, the preference is to use a leading practice rather than best practice, because the term *best* is misleading, as what is best in one scenario might not work in another. Therefore, are there leading practices? Following are leading practices for projects within industries and business sectors (project domains).

33.4.2 What Is the Leading Practice?

The CAMMP™ proposed leading practice is related to a principle suggested early in this book, and it is one of the characteristics of project life cycle and stage gate processes, as explained below.

The stages and stage gates have an important function, which is to provide proper discipline coupled with proper governance, so that the team does not spend time and money on projects and stages that are not yet authorized. The team develops and delivers the project one step (stage) at a time, primarily to avoid overspending at a given stage. When management believes that the project continues to make business sense, and they intend to continue with it, they authorize further work. In other words, in each stage, the team must do what is necessary for that stage to meet the criteria for the upcoming stage gate. Again, only the authorized stage work—no more, no less!

With the above principle stressed, the proposed leading practice is to have at least two main funds approval points and three budgeting steps. This leading practice is common in capital-intensive industries and is applicable in other sectors.

These budgeting steps are at Stage Gate 2, Stage Gate 4, and Stage Gate 5, per the standard CAMMP™ Model.

In each stage, the team must do only what is necessary for that stage to meet the criteria for the upcoming stage gate. No more, no less!

33.4.3 Stage Gates versus Budget Approvals

One point of view is that there is a funds approval point at every stage gate. This statement would be partially correct, because approvals at the earlier gates (SG1, SG2, and SG3) are typically authorizing work, and work is an effort that has associated costs. However, at these earlier stage gates, the authorization is limited to the expenses of the stage work. Some organizations consider the expenses of the initial stages as overhead cost and not formal project capital funds.

The allocation of operating and capital funds, against projects, is typically determined by the organization's financial and accounting rules and is confirmed by the authorized level of management. Capital funds typically require approvals by top executive management or even the board of directors.

33.5 Three Budget Approvals

33.5.1 First Budget Approval

The first budget approval is at Stage Gate 2. This authorization is different from the next two approvals, as will be explained below.

At SG2, the management decision (intent) is to authorize the project to proceed to completion, which is why the sponsor issues the project authorization document (PAD). However, in reality, management is only authorizing the team to continue to the next stage up to its corresponding stage gate. In practice, they are not allocating any funds for capital investment at this time. The funds for the upcoming stage will come from operating budget.

If no money—capital investment—is allocated, then what is the significance of this first budget approval and why use the term *budget*?

Organizations with a sound financial control system and project budgeting process use annual operating plans or business plans that cover the upcoming three to five years. Therefore, when an organization approves the project at SG2, it adds the project and its expected costs to the operating plan or business plan, whichever is appropriate. In other words, it reserves funds for business planning, but these funds are not yet formally authorized or released.

One potential pitfall in this scenario is that some organizations use this budget for the project as the only approved budget (refer to common practices earlier in this chapter). The pitfall is the fact that this "budget" is only a "rough order of magnitude" (ROM), with a wide range. In other words, uncertainty is high, and so are the threats to the project objectives.

33.5.2 Second Budget Approval

The second budget approval point might not be necessary for some projects but is needed for capital investment projects. If it is necessary, then it is common to approve funds specifically for the project, independent of the operating expenses. These would be capital investment funds dedicated to the project independent of day-to-day operations.

Figure 33.1 represents the relevant portion of the CAMMP™ standard model. Note, at Stage Gate 4 (SG4) there is a symbol with the letter "A." This symbol indicates that the team needs to obtain *advance funding* from management to carry out the project definition.

This is called "advance funding" because this approval is not for the final funding.

For capital projects, the level of detail gathered up to this point is not substantial enough to allow for the development of a proper plan with detailed cost and schedule estimates. Consequently, there is not sufficient and accurate information to finalize the economics—financial analysis—of the project. As a result, executive management will not yet be prepared to make the final commitment.

It is worth noting that up to this point the amount of money spent on the project, in comparison to the total expected cost, should be negligible, less than 1–2%. In the event that the organization cancels the project, this low expenditure becomes sunk cost and part of the organizational overhead. In contrast, the Definition Stage is the first stage at which expenditures start to increase, thereby increasing project risk. One of the reasons for the increase is the growing level of resources required, and the fact that it is likely that the project owner will outsource the definition work to a service provider.

If this is not the final approval, at this point what exactly is the team getting?

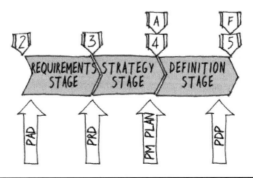

Figure 33.1 Budget approvals across the project life cycle.

The team is getting funds to cover the Definition Stage. A typical advance funding allocation is about 2–5% of the total estimated cost of the project, sometimes even more.[1]

What is required before this approval is the PMP, including updated cost and time estimates. Remember, in CAMMP™ the PMP is based on the Project Requirements Document (PRD) and the delivery and management strategy, which is useful information but not sufficient for final approval.

33.5.3 Third Budget Approval

Figure 33.1 also shows another symbol with the letter "F," representing *final approval*. In other words, final funding normally takes place at Stage Gate 5 (SG5). To obtain this funding, the PDP is a key input.

What is the significance of the PDP?

Organizations with capital investment projects require a detailed plan[2] as the basis (input) for estimating, to produce accurate estimates and reduce uncertainty. These cost and time estimates are accurate enough to allow a major decision to cross the point of no return, at SG5. (Estimating and accuracy are the subjects of the next chapter.)

Cost engineering references show that, at this point, the estimate is within a range of ±5–10%. In other words, the final (actual) project cost could be as low as 90% or as high as 110% of the budget approved at Stage Gate (SG5). From a risk management perspective, organizations typically consider this range—90% to 110%—as an acceptable risk.

Because this approval is a crucial decision point, the request for final approval could take time. Consequently, SG5 could be split into two parts: SG5a to approve the PDP and SG5b to indicate board/management approval of the project investment. In some situations, this may also require financing from external institutions.

33.6 Closing Comments

In this chapter, the focus was on approvals, discussing the various types of approvals in a given project. There are three types of budget approvals.

At the project level, *stage gate approvals* are present at every stage and would be necessary to approve the prior stage work and to authorize the next stage. There are nine stage gates per the standard model, but when tailoring CAMMP™ to a given type, the number of stage gates will vary up or down. The variation in the number of stage gates is also a function of organizational culture.

At the stage level, there are *process gate approvals*. There are four process gates, and CAMMP™ suggests fixing this number. These gates repeat with the processes in every stage.

---

[1] Some organizations are willing to approve up to 10% of the total estimated costs. They do this because they are willing to accept higher risks, and the additional amount allows them to expedite the project and shorten the overall duration. This is a fast-tracking approach.

[2] Also knows as front-end engineering design (FEED), preliminary engineering, or front-end engineering package (FEEP).

Finally, there are the *budget approvals,* which can vary based on organizational preference. These approvals could happen at SG2, SG4, and SG5. CAMMP™ recommendation is to use SG5 if one funding point is required, whereas the preference and leading practice are advance approval at SG4 and final approval at SG5, with initial authorization at SG2.

Chapter 34

Project Estimates

34.1 Introduction

The previous chapter introduced two major funds approval and three budgeting points. The decision-making process at each of the budgeting and funds approvals points is an updated estimate of cost and time. This leads to the following questions:

- What are the leading practices for how many estimates would be necessary for a project?
- Across the project life cycle, is there a need for one, two, or three estimates?
- What about four or five estimates?
- Are the previous statements real? Alternatively, are they complete fantasy?

The Association for the Advancement of Cost Engineers (AACE) International defines five distinct estimate classes for the life cycle of the project.[1] For some domains and in certain situations, the use of five estimate classes, per AACE, is appropriate. Notice that AACE offers a classification system (five classes) and is not necessarily recommending five estimates for every project. The leading practice is three project estimates along the project life cycle (end-to-end), for a project owner organization.

34.2 Stage versus Project Estimates

Once again, it is appropriate to differentiate between *project* estimates and *stage* estimates. Remember that in every stage, the processes repeat, including cost and schedule process steps.

---

[1] The AACE International Recommended Practice No. 17R-97 states, "Five cost estimate classes have been established. While the level of project definition is a continuous spectrum, it was determined from benchmarking industry practices that three to five discrete categories are commonly used. Five categories are established in this guideline as it is easier to simplify by combining categories than it is to arbitrarily split a standard."

Therefore, there are estimates for every stage. How many estimates per stage? Possibly one is enough, but for large and complex projects maybe two, one with the stage management plan (SMP) and one with the stage detailed plan (SDP).

For the project level, the leading practice is to go with three estimates, in a fashion similar to AACE's recommendation.

34.3 Proposed Approach

34.3.1 Overview

What are the three estimates classes?
What is the purpose, accuracy, and timing of each of these estimates?
Are these cost estimates?

Let's start with the last question. The term *estimate* is used to refer to an estimate of the cost and time, unless otherwise noted.

The other two questions are addressed next, with Figure 34.1 showing the placement of these three estimates along the standard CAMMP™ project life cycle. This approach is the proposed leading practice in the application of CAMMP™, even on small to medium projects.

These estimates, including the names, are similar to, but not 100% aligned with, AACE guidelines.

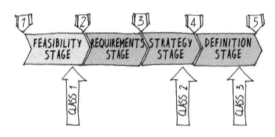

Figure 34.1 Project estimates across the project life cycle.

34.3.2 Class 1 Estimate

The class 1 estimate is a *rough order of magnitude* (ROM) estimate, although *conceptual* or *study estimates* are also common names. This is a high-level estimate, because not enough project details have been developed during the Discovery Phase, when the team develops this estimate. Therefore, it is the least accurate of the three proposed classes.

The primary use of this class is for the feasibility study, in which the team performs the project's financial analysis. For commercial projects, the team focuses on financial return, and for social or governmental projects, on justifying the expenditures in comparison to social or community benefits.

In either case, organizations use this class 1 estimate to make a decision at SG2 on whether to authorize the project or not. Furthermore, as mentioned in the previous chapter, approval at

SG2 leads to the first budget approval and inclusion of the project in the organization business or operating plan.

Project teams typically use analogous or factoring techniques with historical data from previous projects to develop this estimate. Such historical data will typically be at the project level and contain minimal details. In the absence of this information, teams depend on expert judgment and possible quotes from third parties, such as vendors and service providers, although this is not always practical or appropriate for a given project.

The accuracy of this estimate could be ±50% or some other range. Therefore, this estimate is not accurate enough to control the project, but it is sufficient (within the range) for management to decide if the project is feasible.

Keep in mind that the term *estimate* refers to both cost and time. Therefore, the class 1 estimate is an estimate of the approximate project cost and expected project duration. However, concerning duration, it is not necessarily ±50%, although it can be. Schedules are often linked to imposed or preferred deadlines, which in some cases might not be realistic. For the class 1 estimate, the schedule information is only at the summary level.

34.3.3 Class 2 Estimate

This second estimate is an update to the class 1 estimate and might be called the *planning estimate* or *budgetary estimate*. The team develops this estimate during the Strategy Stage, as a part of the project management plan (PMP).

The main purpose of this estimate is the enhancement of management's understanding of the project's cost and duration, to determine whether it is still in line with expectations and whether the project should continue beyond Stage Gate 4 (SG4).

The accuracy of this estimate is higher than the class 1 estimate, possibly by ±25%. In other words, the estimate is still not accurate enough for proper (quantitative) control, but it fits the purpose of updating the feasibility of the project and gaining approval at SG4, with advance funding, if required.

The project team typically uses a blend of historical data from previous projects and other estimating techniques to develop this estimate. This approach varies within industries and companies and is a function of the project management maturity of the organization.

34.3.4 Class 3 Estimate

This is a critical estimate, which the project team develops during the Project Definition Stage. This is called the *detailed estimate, semi-definitive estimate,* or *funding estimate*. One may also encounter the term *control estimate,* because this estimate is used after approval to develop the performance management baseline. The project team then uses this for project control purposes during the Implementation Stage.

The project team uses the information from the Definition Stage plus the project planning information, including the detailed work and product breakdown structures, to develop the class 3 estimate. With this level of detail, the estimate is sufficiently accurate (±10% or even narrower) and is suitable as the basis for the final project approval at Stage Gate 5.

34.3.5 Summary of the Three Estimates

In summary:

- Each estimate has a different purpose.
- The purpose is appropriate to a related stage in the project life cycle.
- Each estimate is input to a stage gate related to budget approvals.
- The level of estimating details and estimate accuracy is a function of the Project Stage.
- As a result, the uncertainty of the estimates is reduced with the subsequent estimates.
- Reducing uncertainty leads to reducing risk to an acceptable range.
- Reducing threats will enhance the chance of project success.
- Finally, an accurate estimate provides proper and realistic control baselines.

34.4 Are Three Estimates Too Many?

Maybe, maybe not!

Various resources and studies still reflect that a significant percent of projects fail, or at least are completed with less than the optimal outcome and expected benefits. Many are behind schedule or over budget. Even if the project is within the cost and schedule parameters, that is not necessarily a good sign if the team had inflated the project targets.

Why such a poor record?

It is not necessarily that poor estimating causes failed and challenged projects, but it would be fair to ask, "Is estimating a factor?"

It would be appropriate to expand the discussion from estimating to the use of these estimates in making project decisions. CAMMP™ links estimating directly to the stage gates. Let us pose the question again but in a different way: "Did these failed or challenged projects follow a disciplined approach, using techniques and controls such as project life cycle and stage gates?"

The issue here is whether an organization's management wants to approve projects earlier in the project life cycle and with less detailed estimates. Let's say they want to approve projects using a class 1 estimate as the only basis for funds approval; then they must expect a higher risk of project delays and budget variances. This consequence exists because the level of detail for the class 1 estimate is low, and the accuracy is the least of the three estimates. Unless these estimates are grossly inflated, what is the chance of success?

In other words, there is a tradeoff between earlier decision making and higher risk. If this tradeoff is acceptable, then great, no issues. If not, think twice about this approach and consider adopting the proposed leading practice.

34.5 Components of a Good Estimate

What are the elements of a good estimate?

Regardless of the estimate classification, each estimate should include the following costs. However, estimating them could vary from one estimate to another. Estimating techniques could be based on factoring, extrapolating, analogous, or semi-detailed approaches.

- **Direct cost.** This is the cost of the labor and materials needed to perform the work of the project. Materials refer to the things that will be installed, and labor cost is limited to the physical work done to install facilities or perform other technical or functional work.
- **Indirect cost.** This is the cost of direct supervision and associated costs, such as offices, equipment, and other resources that are required to perform the work.
- **Management cost.** This is the cost of the management resources that are directly working on the project—the project management team.
- **Overhead cost.** In some organizations, there are costs related to organizational management that is not working directly on a given project. In project-driven organizations, this cost is usually a small percent and applies to all projects.
- **Inflation cost.** Because some projects will require more than one year's duration, the actual expenditures could extend a year or more from the time of the estimate. Therefore, the estimate should include escalations to bring the cost from the estimating time to the expected time of expenditure, per expected inflation.
- **Cost of money.** This applies to large projects and includes the cost to fund a project in terms of interest rates and financing fees. If the project has global procurement, another item to consider would be currency exchange cost.
- **New technology allowance.** If the project has components based on new technology that is not commercially proven yet or the team is not familiar with, a special allowance could be of value both to deal with the potential bugs and deficiencies and to apply to the learning curve.
- **Contingency.** This is an allowance to deal with the potential threats on the project that can be estimated or expected. Some organizations calculate this amount based on the output of the risk management processes; others apply a fixed allowance that could be 25–50% for a class 1 estimate and 5–10% for a class 3 estimate.

Usually, all of the above costs will be part of the project budget that is under the project manager's control.

- **Management reserve.** This is the final allowance that the estimate includes on top of the estimated budget amount. The sponsor or a steering committee typically manages this allowance. In our experience, this is a financial control and governance rule that allocates about 10% above the project budget controlled by the project manager.

34.6 Estimating and Project Management Maturity

There is a link between proper estimating and project management maturity. In general, the link is not limited to estimating, because all project management processes, particularly planning, are a function of an organization's project management maturity. However, this book is not about project management maturity and will not explain the links in detail but will touch on the most important points.

The most important factors that contribute to a higher level of project management maturity are components of a formal organizational project management system (OPMS), such as methodology, governance, processes, lessons learned, historical data, and project records.

The historical data and project records have a direct impact on estimating and estimate accuracy, because the estimating effort will use proper records and actual data instead of guesswork or vendor quotations. Furthermore, vendor quotes may not be possible during the early stages of the project, because vendors need detailed designs or specifications.

A non-existent or weak organizational system, lack of historical data for cost and time, and an absence of lessons learned will all contribute to poor planning, unreliable project detailed plans (PDPs), and funding decisions based on questionable estimates. Moving on to implementation, poor planning will lead to many variances and unanticipated risks, which in turn result in the project management team's spending too much time reacting to problems instead of practicing proactive management. In these situations, the outcome is almost certain—failed or challenged projects.

The above only touches on the importance of proper estimating and project approvals and their role in effective management of projects, and, in turn, higher organizational project management maturity.

34.7 Estimating and the Project Life cycle

Figure 34.2 presents a comparison of the three classes of estimates and final cost as compared with the level of project scope definition.

In this image, the dark shade represents the value of the defined scope—what is known, cost or time.

The true (actual) final cost cannot possibly be known with certainty until the project is complete. However, project teams are challenged to approximate that final cost early, as soon as the feasibility study.

How could they do that?

One choice is magic; another is guesswork; still another is to get information from vendors, assuming that is possible.

However, the leading practice is a factor of project management maturity, as discussed above. If the organization has the right level of maturity, lessons learned, historical data, and project

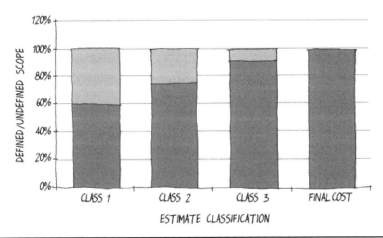

Figure 34.2 Estimate classification and level of scope definition.

records, and the results are broken down by project type and estimate class, then it would be possible to estimate per the appropriate accuracy.

In Figure 34.2, for the class 1 estimate, which the team develops during the Feasibility Stage, the team only knows certain things about the project—at a high-level, not detailed. Using historical data, they can estimate what is known, shown in dark shading. In this image, the 60% representing defined scope is an arbitrary number; it could be 50% or 70%. It is evident that if the team were to use this amount and approve the project on that basis, they would have grossly under-estimated the cost of the project, and the project would fail or at least come in significantly over budget.

What is the solution?

Back to maturity and history. With good history, the team can determine that for a typical class 1 estimate, they would need to apply about a 25% (or even 30% or more) contingency to approximate the final cost of the project, and they would apply this contingency to cover the unknowns (light shading).

For the class 2 estimate, the team would have progressed the project, and they would have finalized the project requirements and most of the PMP. Therefore, they would know more than at the previous stage, hence the larger dark area. Consequently, the team needs less of a contingency to approximate the final cost.

The same analogy repeats for the class 3 estimate. The team knows much more after developing the PDP, but not everything. In this case, the team might know 90% or more,[2] so they still need about a 10% allowance to approximate final cost.

To reach project management maturity, organizations must save all of the estimating data and estimates in the historical database. In the Close Stage, the team will reflect on these estimates and perform the reconciliation between them. The team might find that the contingency used was either not enough or too much, and they need to understand and document why. A summary of the final reconciliation report would be uploaded to the project management information system and update the project's records for future use.

If organizational management or the project management department notices a trend or repeating gaps in one place or another, that might be a trigger to modify some processes and procedures.

It is a continual learning process.

34.8 Closing Comments

How many estimates will an organization use on their projects? What is the purpose and accuracy of each of those estimates?

All of these questions and more are a function of the organizational culture, level of maturity, and project type or classification. Another factor is the organization's tolerance for risks. The recommendations and the CAMMP™ approach consider three project estimates and at least one estimate for each stage. However, this would be one of the things that would be tailored to organizations as they build their comprehensive organizational project management system.

---

[2] Although the figures used could be realistic, they should not be used as guides, because they would vary either from one organization to another and among the functions of project type, domain, and class.

Chapter 35

Project Control

35.1 Introduction

It should be clear by now that control, like all other functions, is at two levels: the project level following the project life cycle, and the stage level following the processes.

Furthermore, the concept of *baselines* is established in project management; however, in this chapter, the elaboration provided will offer a slightly different perspective.

35.2 Baseline

The common definition of a baseline is *a line of reference, the basis against which we compare project performance.* In other words, the baseline is essential for project control, because it reflects the "approved plan." Then the team performs project control by comparing actual performance against approved plans.

Where is this line of reference?

What plan should the team use for control and measuring performance?

As stated repeatedly, the traditional view, or common practice, is that project control is about comparing actual performance to the plan, which is typically the one used to obtain approval (the approved plan), which means it is the project management plan per the *PMBOK® Guide* or ISO 21500. However, it is vital to expand on this traditional view.

35.3 Stage versus Project Control

35.3.1 Stage Control

If one considers the CAMMP™ processes, then control starts as soon as the team starts the Authorize process, with the control reference being the statement of work. Furthermore, control

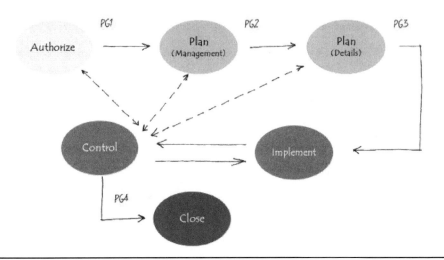

Figure 35.1 Stage control and stage control reference points.

continues during all of the other processes. This approach is the concept that is implied, but not clearly stated, in the PMI and ISO guides.[1]

Shifting to CAMMP™, Figure 35.1 (repeated from Chapter 16), shows the six processes. Notice that the Control process is linked to the other five processes via the dashed or solid lines. These lines are to indicate that control is active throughout the stage.

As the work proceeds from one process to another, the control reference point changes as it is superseded by the new control reference. For example, during the Authorize process, the control reference is per the idea statement or project brief. Once the stage authorization is issued and approved at PG1 (Process Gate 1), that document will become the control reference during the Plan Management process work. In other words, as the project management team develops the stage management plan (SMP), they must continue to compare to their control reference, the stage authorization document.

Similarly, once the SMP is published, it will become the control reference point during Plan Details process. Next, during Implement process, the control reference for validating the work is the stage detailed plan (SDP). Finally, the output verification for the stage would be at PG4, just before closure.

35.3.2 Project Control

The CAMMP™ approach for control on the project level is next.

---

[1] We will continue to use CAMMP™ terminology, with *process* referring to one of the six processes of Authorize, Plan Management, Plan Details, Implement, Control, and Close. We will also use *process steps*, or *steps*, to refer to any step under these processes. For example, Establish Stage Success Criteria is a process step within the Authorize process of a given stage.

35.4 CAMMP™ Alternative Perspective

35.4.1 Overview

Many organizations that have a good level of project management maturity use one form or another of the concepts that are presented in this section, but possibly *implicitly*, not necessarily *explicitly*.

The proposed approach is that the baseline is not fixed, and it is not limited to the project management plan (PMP). The baseline moves as the project progresses from one stage to another (as shown in Figure 35.2). The shifting baselines, or what CAMMP™ calls *control reference points*, start with the project brief.

It is important to note that *moving baselines* does not mean *changing the baseline*. In this context, changing the baseline refers to the act of replanning the project and changing the approved plan's target of cost and schedule. Unless the project objectives change, changing the baseline is not a proper project management practice. However, the term *moving baselines* means that, as the project progresses from one stage to another, a new control reference point will supersede the previous one.

Well, is this not an example of changing the baseline?

Not really—read on!

35.4.2 First Control Reference Point

In Figure 35.2, the lower arrow touching the project brief on the left represents control in comparison to the project brief, the idea, and business case of the project. This is the first control reference point.

Can a team control against the idea?

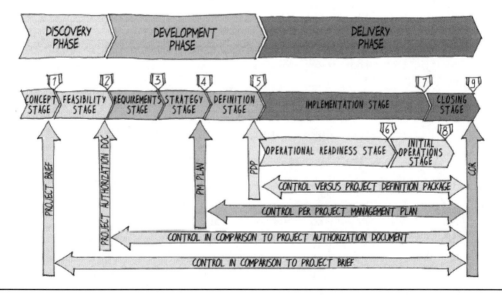

Figure 35.2 Project control with the concept of *moving baseline* (control reference points).

Yes, they can, although control in comparison to the idea is conceptual and not quantitative. The phrasing of the project brief is critical, and a slight variation can indicate different projects. For example:

- **Scenario 1.** The project's purpose is "to get drinkable water from city A to city B."
- **Scenario 2.** The project's purpose is "to build a pipeline to get drinkable water from city A to city B."

Clearly, Scenario 2 is more specific, and one would expect that the project manager will work on building the pipeline. On the other hand, in Scenario 1, a good project manager will look at all options for moving water and choose the optimal one.

To build on the above, if the project manager in Scenario 2 starts considering options other than building the pipeline, then he or she is deviating from the project mandate, which would be outside the project objectives. The role of project control is to stop or eliminate such deviation.

However, in Scenario 1, if the project manager is not evaluating the various possible options for moving water between these cities, then he or she would not be performing per the project mandate. Meaning, the project manager would be deviating from the project objectives.

Therefore, the highest-level control on the project, during the first phase, is working in accordance with the project brief. In other words, this initial control action is to ensure that the feasibility study is developed per the project brief and does not deviate.

35.4.3 Second Control Reference Point

Control is an ongoing effort that starts with the project brief, as already stated.

The second control reference point is in comparison to the project authorization document (PAD), which is the second arrow from the bottom of the figure. Assuming the team has performed a proper feasibility study, and the sponsor has developed a clear PAD, then the second control reference point supersedes the first, because it will encompass its purpose. Stated differently, the PAD incorporates the project brief and expands on it.

One may ask a similar question: "How can the team control against a brief, high-level document, such as the PAD? There is not enough information to control against at this level; there is no baseline yet."

The proper response is:
"There is enough to exercise control!"

The team has the project description, justification, and objectives—all part of the PAD—and these are the ultimate control points. They also have a rough-order-of-magnitude estimate.

Aren't these items vague?

Agreed, they do lack detail, but if the PAD's brief contents cannot clearly define the project objectives for the team, then it has a deficiency, and there is an issue that is likely to impact the project's performance. Therefore, control must continue as soon as the sponsor issues the PAD.

The team might not be able to use the full, earned value management approach at this point, because they do not have quantifiable means for control of the entire project, but that is perfectly acceptable. The main concern during the Development Phase is to ensure that the team is developing the project per the project objectives as defined in the PAD.

What happens if there is a need to change the objectives?

This can be done via two mechanisms:

1. The sponsor can amend the PAD, or
2. The organization should have a *project change control system*[2] that includes the policy and guidelines on how to initiate and approve changes to the PAD. These are project objective changes.

The second approach is recommended, because it would document a change for historical purposes and project reconciliation.

One final thought: The previous chapter defined the various estimate classes. It also noted that the class 1 estimate is a key requirement for the feasibility study. As a result, this second control reference point has scope, time, and cost components and can essentially be a preliminary baseline appropriate for the level of detail of the PAD.

35.4.4 Third Control Reference Point

So where is the third control reference point?

How about the project detailed plan (PDP)?

Not yet—before the PDP there is the PMP.

After the PAD, the project manager and the team start working on the project requirements in line with the project objectives. Subsequently, the team develops the PMP. During these stages, the project manager must use the PAD as the guiding reference and cannot deviate from it without the formal approval of the sponsor. The project requirements document (PRD) and PMP must be complete per the PAD. *Once the PMP is ready, which includes the updated class 2 estimate, the team has the third control reference point or baseline.*

The PMP includes various project management elements, such as procurement strategy, project life span, updated schedule, cost estimate, etc. These are more than enough to allow us to exercise control *during the Definition Stage*. The team must monitor and control for any change or variation from the PMP.

35.4.5 Fourth Control Reference Point

What is the fourth control reference point?

This fourth point is easy! It is the traditional set of control baselines. These are the baselines established by the PDP, which the team will use extensively during the Delivery Phase.

35.4.6 Why Four Points?

It is vital to note here that the team is not necessarily, and concurrently, controlling against these four control reference points. In practice, these control reference points will supersede each other.

Why is there a need for four points of control?

Is this approach necessary?

---

[2] Part of the overall organizational project management system.

Is it not enough to control against the baselines we defined per the PDP?

To answer these questions, take a step back and reverse the question, "Is one point sufficient?"

One reason that the traditional and shared view of control starts with the PDP is its linkage to project funding at final approval. Some practitioners think that unless there is an approved budget, there is no need for control. In other words, they link control to the money in an approved budget.

In other cases, there is no formal control or change management unless there is a contract for services, because the view is that internal changes are part of the development and so are not true changes—changes only exist in the context of a contractual relationship.

Project control is not only about money or contracts. Project control is about variations from the plan, including variations in scope, time, cost, quality, and human resources.

Notice the use of the word *plan* and not *project detailed plan*: The plan in question is any official reference (the last approved deliverable or estimate) that the team can compare against, such as the PDP, the PMP, and even the PAD. Therefore, the team must monitor for variations and changes throughout the project.

35.5 Project Management Maturity and Control

OK, clear, but still, what is the significance of project control during the Development Phase: wouldn't the PDP absorb all of these variations and changes and become the document that supersedes all before it?

Well yes, but is that enough?

Let us put on the organizational maturity hat.

The assumption here is that during the Development Phase, the team would be expanding the level of detail in the project scope from the Discovery Phase. This development of concept action—progressive elaboration—will likely introduce details that were not listed earlier. Some of these could be deviations, while others could be changes or scope creep.

By "absorbing" these variations and changes in the PDP without a formal project control and approval mechanism, the team is essentially burying or hiding these changes in the project details. If these variations and changes are due to poor planning, lack of experience, or an inadequate project management system, then how can the organization know, how can they learn, and how can they seek continual improvement?

Alternatively, the changes could be due to environmental factors, changing economic conditions, or numerous other factors. It does not matter what the reasons for variations and changes are—the team must know them and understand them. They also must document them, so that the organization can learn from them and build toward higher levels of project management maturity.

Chapter 36
Project Change Management

36.1 Introduction

Project change management and project control have a similar foundation. The four control reference points discussed earlier apply to change management and are equally important.

There are numerous resources and references on project control and project change management. Therefore, we will continue with the practice of referring the readers to research these topics for more in-depth coverage, while this book maintains its focus on these subjects as they relate to the Customizable and Adaptable Methodology for Managing Projects™ (CAMMP™).

> Project control is about variations and deviations from the plan,
> which happens as a result of performance-related matters (doing work).
> Whereas (project) change management is about the conscious
> decisions to change the plan, or part of the plan!

Although it might be clear to many readers, it is of value to differentiate between control and change management. Control is about *variations and deviations from the plan stemming from performance-related matters,* internal or external, individually or organizationally caused. On the other hand, *change management is about the conscious decision to modify the plan.*

36.2 Agile/Scrum

Before moving on and shocking Agile/Scrum users, it is vital to have the proper perspective.

Agile and Scrum might build on concepts of flexibility with limited up-front planning, then planning, designing, and building in increments or sprints. That might be an acceptable approach for software development, although this is often under debate.

Outside of software development, project management has to follow a certain structure, with a project life cycle and phases/stages. For these types of projects, proper change management is vital for success.

36.3 Why Change Management Is Critical

For projects outside the software development domain, change management can be critical and a major contributor to project success or failure. In general, changes to projects are disruptive and are leading causes of threats to the project's objectives. Studies by the Construction Industry Institute (CII)[1] for capital-intensive projects—industrial and engineering/construction projects—find that poor change management results in poor productivity, and in turn over-budget and behind-schedule performance. On the other hand, effective change management can lead to better-than-planned performance.

If the team does not adequately manage changes, starting with proper identification, documentation, and analysis before decision making and implementation, that is a form of weak project governance. This lack of change management is a critical factor for organizations without a proper project management system and an absence of proper accountability.

CII has a best practice related to change management, and the following are some of its highlights and some of the issues and threats from project changes:

- Impact to engineering and construction productivity due to high level of changes.
- Impact to project performance due to changes.
- Impact due to hidden changes.
- Organizations fail to identify the true impact of changes.
- There is a cumulative impact of changes, which the team cannot readily evaluate!

All of these are factors that confirm the need for proper change management, especially in the "foggy zone," from idea to SG5 (refer to Section 36.5, Change Management in the Fog). The remainder of this chapter focuses on three areas of attention about project change:

1. **Hidden changes.** These are changes that do not go through the project change management system, assuming there is one.
2. **Project change management in the fog.** This is a term the author uses to refer to the part of the project life cycle during which control is mostly qualitative owing to the ambiguity of the scope definition.
3. **Project change management during implementation.** This is the most common in traditional project management.

36.4 The Hidden Changes – Scope Creep

36.4.1 Overview

One of the key advantages of the concept of four control reference points addressed in the previous chapter is to avoid *scope creep* (creep, as in *sneak in, creepy, sinister,* or whatever term the reader prefers to use); this is a major challenge in projects.

---

[1] http://www.construction-institute.org

Why does scope creep happen?

One factor is that the project brief and project authorization document (PAD) are high level, lack detail, and therefore leave room for interpretation. Even the project requirements document (PRD) and project management plan (PMP) do not have the same level of detail as the project detailed plan (PDP).

Another factor is that, as the project progresses along the project life cycle, the level of detail increases.

As a result of the above factors, it is often difficult to decide what is (1) a natural variation of scope, (2) a conscious change, or (3) a natural expansion of the project details owing to progressive elaboration. Although it is difficult, it is possible, with proper due diligence, to tell the difference between a variation, a change, and design growth.

Time for an example. This example focuses on an air conditioning system for a villa.

36.4.2 Discovery Phase

The project is to build a large villa, with a three-car garage and a garden, among other features. The villa will have air conditioning and heating, but the garage will not. The project is feasible, and management authorizes it.

36.4.3 Requirements/Strategy Stages

During these stages, the architect/engineer works on the concept design and determines the approximate size of the air conditioning unit as size X. Its cost is included in the class 2 estimate.

36.4.4 Definition Stage

As the project moves into detailed planning, the team will have a better idea about the house size, details, windows, and insulation. The team must revisit the air conditioning size and determines that it should be a size X1, which is about 10% growth over the size defined earlier. *This is normal design growth stemming from design development* (progressive elaboration). Based on the detailed design, the team updates the budget in the class 3 estimate to cover size X1.

When the team investigates availability in the market, they find that there are only sizes X, X2, and larger. X is too small, and X2 is greater than required. What does the team do? The team will have to undersize or oversize, and in either case, there is *a variation from the plan. This growth is a market-driven variance.*

So far, what is the situation? In comparing the current situation with the earlier basis:

- There was a 10% growth owing to design development, from X to X1, assuming X1 is 10% over X; and
- There was another 10% growth stemming from market conditions, assuming X2 is 20% over X.

These are normal and expected variations. If the team has kept records from past projects, then they might even have expected these variations and included them in the estimate—normally part of contingency. While estimating, the team might not have known the exact variations or magnitude, but they know these things happen—"the known unknowns."

As the work progresses, still in the Definition Stage, the owner decides to air condition the garage, resulting in an increase in the unit size to X5, a 50% growth over the size defined in the requirements. *This increase would be a change to the objective.*

In this case, the market variance is not relevant anymore, and the growth caused by the change is from (the unavailable) size X1 to size X5. In other words, there is 10% growth due to normal design development (normal variation) and 40% due to a conscious decision to change the design.

Notice that if the team has adopted the traditional view of control, which is control per the detailed plan, then the air conditioning unit would be updated to size X5 in the PDP. Consequently, the class 3 estimate needs to include a budget allocation for this larger size. This growth is 50% over the class 2 estimate. Then when this project is over, the size X design would be forgotten.

If the team had not kept records and did not impose the project change management system during project definition, they would have missed this change. In other words, this growth would have been hidden in the PDP.

No big deal?

It *is* a big deal.

The air conditioning unit increased in size from X to X5, a 50% growth, and only 10% of that was due to the normal progression of design (from X to X1). This increase in size had a corresponding increase in cost and possibly schedule and risk, among other factors. In other words, the actual magnitude of the change would have gone unnoticed.

Now extrapolate this situation to the windows and doors of the house, the landscape and irrigation system, roofing and flooring, what would happen? How about making the house bigger or smaller? How can the organization distinguish design development from changes?

36.4.5 Why Scope Creep Happens

Often, scope creep could be innocent, and things happen as a result of stakeholders' discovering missing items, so they add them during the Definition Stage. Missing things is often a sign of a poor system or lack of proper planning. However, assume these are innocent mistakes.

However, scope creep is often not innocent. Stakeholders use the ambiguity of the earlier stages to add scope—preference items—for things that would not be required to meet the project's objectives. These are hidden changes that get added without proper justification or documentation and might not be discovered until the project cost in the class 3 estimate shows significant growth.

36.5 Change Management in the Fog

Project change management in the fog is a term the author uses to refer to the part of the project life cycle during which control is mostly qualitative because of the ambiguity of the scope definition.

Why use the analogy of fog?

Well, imagine you are driving in extremely dense fog; you would not be able to see anything, and driving would be dangerous, so you should stop. Relating this concept to project management, a project in which the team does not know the scope *is not a project and should be stopped.*

Now, imagine that you are driving in dense fog; you will be able to see some things, but not clearly enough. Driving under these conditions is dangerous, but if one must drive, extra care is in order. Even under those circumstances, one would be able to see a few meters ahead—not the details, but enough to stay on the road. This would be similar to the time span between project brief and PAD, during which the team does not see the details but would know if they sidetrack off the project objectives—assuming the project brief is clear and unambiguous.

Continuing with the above analogy, imagine you are driving in fog but not as dense as before. You will be able to see things, not very clearly, but enough to pick up speed while maintaining caution. This would be similar to the time span between PAD and PMP or between PMP and PDP, depending on the density. In this period, there is more clarity but still not enough to see the full details; that is, not quantitative, such as the air conditioning example.

Why the above analogy?

Because of the traditional view of project change management, some organizations do not implement a change management process until after full funding, from SG5 onward, and in some cases only in relation to contracts. To these organizations, unless everything is clear, quantitatively, there is no control, which is not acceptable. As stated earlier, the early stages, when things are not clear, is when scope creeps into the project and might not be discovered until too late. Hence, the need for careful and proper monitoring and control.

36.6 Traditional Change Management

Traditional project change management refers to the common practices for managing project changes. In many organizations, management of the project's changes is limited to projects for which the work is under contracts, and the changes would be specific to the contracts. This approach represents control against the fourth control reference point, as discussed in the prior chapter.

With this approach, the team is skipping proper control between control reference points 1 and 4. This period is the foggy, ambiguous area—the critical area.

36.7 Types of Changes

There are different kinds of changes that happen on projects.

- There are common contract changes, when working under a contract.
- There are project changes that do not alter the project objectives. These could be changes in the plan or other changes. These changes could happen whether there is a contract or not. Normally, these changes are within the control of the project manager and are covered by the project budget.
- There are objective changes. These are defined as any change to the project objectives outside the control of the project manager. For example, moving the location of a facility project, altering the capacity of a production facility, adding features to a software project or a new product. Usually, these types of changes will require supplemental funds and extension of time, because they are outside the power and influence of the project manager.

36.8 Closing Comments

In the previous discussions, specifically on scope creep in the air conditioning example, one can notice:

- Natural growth caused by design development (progressive elaboration)
- Growth stemming from market conditions, which became irrelevant later
- Growth brought about by changing requirements

This information is crucial for learning about what happens on a project and is included in the reconciliation report during the stage closures and the Close Stage. This is also necessary for organizational project management maturity, as discussed more than once. Although this was a simplistic example, it is a reality.

Repeatedly, there are reports on projects that went up in cost 10, 20, 30% or even more between the class 2 and class 3 estimates, and management is asking why.

"It's the market conditions!" is a common answer.

Well, maybe it is the market conditions. However, the market conditions in normal times do vary, so if something goes up usually something else goes down and balances it.

Then what are the reasons and the threats to the organization?

Perhaps, during the Discovery Phase and feasibility, the organization determines that the project potential returns are low but still wants to consider the opportunity and take the risk. However, if during the Development Phase the project cost jumps by 10, 20, or 30%, as discussed above, then the opportunity will not be feasible anymore and the organization is likely to stop the project. This scenario leads to two questions:

1. Did the organization just kill a project that would have been beneficial if it were not for the hidden changes and scope creep that took place during project development?
2. How about all of the wasted effort related to the work done until the point of cancellation?

What would be worse is for organizations not to see the negative signs and continue with a doomed project.

We leave you with a few points to ponder:

- Do you see a need for project control during the Project Development Stage?
- Can you reflect on your projects and think about potential changes and scope creep that affected your project outcomes?
- Could you have avoided a cost or schedule overrun if you had applied these principles?

Chapter 37

Project Risk Management

37.1 Introduction

Project risk management, as a general topic, is outside the scope of this work. For further reading on this subject, there are many publications from ISO (ISO 31000), PMI, and other organizations and individuals. The inclusion of this chapter is to highlight potential gaps in the practice of project risk management, particularly gaps related to managing risks across the project life cycle.

37.2 Threats or Threats and Opportunities?

One of the shortcomings of the practice of project risk management is unique to the concept of risk perception. There are two schools of thought on risks:

- One considers risks as threats to the project. As a result, this school of thought uses the term *risk management* to refer to the management of threats.
- The other school of thought considers risks as threats and opportunities, meaning a risk event could have positive or negative consequences on objectives.

The author subscribes to the latter school of thought. A risk is an event; if it happens, its implications could be a threat or an opportunity, and risk management is about managing risks by minimizing the threats and maximizing the opportunities.

If an organization prefers to consider risks as threats only, that is a choice. However, in that case, these organizations must manage both threats and opportunities.

37.3 Frequency of Performing Risk Assessment

How often does your organization conduct a risk assessment exercise on a project?

A common response to this question is, "We do risk assessment once during planning, but we monitor throughout the project."

Although this is a common practice, it should not be a leading practice. Furthermore, this answer reflects a gap, which relates to the confusion between process groups and project life cycle. Some practitioners think that the *PMBOK® Guide*'s planning process group is *a Project Stage—the Planning Stage*. Because they believe that planning is a stage, then risk management assessment, which is part of the planning processes, takes place only during this stage. However, because the process groups and their planning processes repeat in every stage, then risk assessments and the corresponding risk response planning must also repeat in every stage throughout the project life cycle.

Please note that the above is different from risk monitoring and controlling. Risk assessment means risk identification, analysis, and prioritization. With that done, the team will take the necessary actions to deal with high-priority risks and develop response strategies.

In closing, risk assessments and response planning take place during the planning processes of every stage. Then during the implementation of the stage, the team would monitor and control risks on the watch list.

37.4 Stage versus Project Risks

The previous section introduces another challenge.

With every project, there are different types of risks, such as business, marketing, safety, planning, and much more. However, in the context of the project life cycle, risks are divided into two categories.

The first category is the set of risks that are unique to a *stage,* meaning they will not be relevant when the stage is complete. Therefore, the team must treat these risks during the given stage or lose the opportunity for effective risk management. For example, risks that could affect the development of the project feasibility study will not be relevant once the study is completed and accepted. The same applies to the other stages. These risks are addressed during the stage detailed planning work.

The second category is the set of *project* risks; these are risks that affect the project's overall objectives and are not particular to a given stage.

37.5 Risk Management Across the Stages

As early as the feasibility study, risk management is a topic to address and revisit frequently.

37.5.1 Discovery Phase

The Project as an Opportunity

Let's look at a project from the risk-management viewpoint.

What is a project?

A project is an opportunity that the organization is considering, such as a new product, a new market, or a new revenue-generation idea. A project could also be a response to a threat, such as a competitive threat, regulations, or safety/security matters.

To elaborate, we will consider that the project is for a new product—a business opportunity—because if the organization delivers the project successfully, it will create the capabilities that would enable the realization of benefits, or positive consequences. If the team fails in delivering, the consequences will be negative.

Within the above context, a feasibility study is a form of risk analysis in which the organization assesses the risk event (the project/opportunity) and determines whether the benefits justify pursuing the possibility. Through the feasibility study, the organization decides whether to exploit the opportunity presented to them.

Project Risks

During the development of the feasibility study, the focus is on commercial risks, market risks, and financial risks, among other high-level, project-wide risks. These are risks that the team should address during the work on the feasibility study. If management considers that the threats from these various risks are significant, the organization would likely decide not to exploit the opportunity and would discontinue the project.

However, let's say that at Stage Gate 2 (SG2), executive management decides to go ahead with the project; in that case, they are indicating that those risks (threats and opportunities) are manageable, and the team has considered those risks as part of the project's economic calculations. In other words, they have either managed or accepted the risks. Alternatively, they could have accepted them at the time of authorization, on the condition that the team address those risks when appropriate, in the upcoming stage(s).

Stage Risks

As mentioned earlier, in addition to project risks, there are stage-specific risks that the team must deal with during the planning for the Feasibility Stage, or they may affect the product of the stage.

- Some of these risks can be people related, such as human resource availability and competence.
- Other risks can be time related—for instance, not enough time to do a proper study.
- Other risks can be a result of lack of organizational systems and standards.
- Then there are the possibilities of poor planning and known distractions during the implementation of the stage work.

Because these are risks that only affect a stage, once the stage is over these risks are gone, whether they materialized or not, unless they are treated and managed. How does the team manage these risks? Like any other risk, the team has to identify them, analyze them, and respond to them.

Why is this important?

Consider the following scenario. Management makes a decision to spend millions of dollars, if not more, when they approve a project based on the output of the feasibility study. Such a decision builds on the understanding that the feasibility study and conclusion is good. Can one imagine the consequences if the feasibility study conclusion proves to be faulty? If it was based on poor analysis or by personnel who did not have the right level of competence? These are risks specific to the Feasibility Stage and, if not addressed, then all else is questionable, and the decision to approve the project lacks the proper due diligence.

A suggested resource for further study of this topic would be ISO 31000, which is a guide for enterprise risk management.

37.5.2 Other Project Stages

For each stage, the team faces:

- Project risks that they carried over from the prior stage
- Project risks that they identify in the given stage
- Stage-specific risks for the given stage

Regardless of the risks' categorization, the team must manage them all.

As the team moves ahead with the project and develops the project authorization document (PAD), project management plan (PMP), and subsequent deliverables, they will start to deal with different types of project risks—the project development and delivery risks. This cycle repeats until the project is complete. It is important to note that, from an organizational perspective, there are risks even during the Close Stage, although these risks will not affect the specific project objectives, since it is complete, but will impact organizational learning, specifically for project and risk management.

37.5.3 Timing for Managing Risks

Just to be clear and avoid misunderstandings, what is presented here does not mean it is acceptable to move risks from one stage to another. The team must pay attention not to defer treating the risks into later stages if they can address them during the stage at which they were identified. *The project team must have the proper discipline and due diligence not to defer risk treatment.*

For some risks that the team might identify early, they might not have enough information to deal with these risks at the time of identification. In that case, they can defer their management to the first upcoming stage in which the team can manage them effectively.

Chapter 38
The People Aspects

38.1 The Project Stakeholders

In general, project stakeholders have different involvement or influence on the project, and one might group them into three categories:

- The project team—the people directly involved in the project
- Those stakeholders who have direct influence on the project or are affected directly by the project
- Stakeholders who might be affected by, but have no direct impact on, the project

To manage the relations effectively, it is important to realize the differences among these three groups. Another possible categorization of stakeholders is relative to their positions toward the project: are they supporters or detractors of the project, and is their support (or opposition) mild or extreme?

Once again, we refer the reader to research other resources to explore this topic further.

38.2 Stakeholders Across the Project Stages

Is there one set of stakeholders (and team) on the project from start to finish?

Are the stakeholders the same throughout the project life cycle?

The answers to these questions vary from one project to another and one domain to another.

It would not be practical to address every type of project here, so here is a general view that would be common across projects and domains.

In the bus trip analogy below, the focus is on those stakeholders who are directly involved with the project. Although there are other stakeholders, it is not practical to include them in this discussion, because how they are affected is highly sensitive to the project domain.

38.3 The Bus Trip Analogy

A good way to explain the key roles on the project is to imagine a bus, with the stakeholders on the bus taking a trip, the trip representing the project.

The driver of the bus might not be the ultimate decision maker, but let us agree that whoever is driving the bus has full authority, or at least the necessary powers to make some decisions.

Please join us on this trip!

- A project starts with an idea. Therefore, there must be an idea originator.
- The idea originator proposes the idea to management. If management accepts the idea, management appoints a sponsor. In other words, if management accepts the idea, the organization hires a bus and puts the sponsor behind the wheel.
- It would be good to have the project manager on the bus, or at least a representative of project management. Remember, CAMMP™ advocates that a project manager must lead the Discovery Phase—the "pre-project" work.
- At the starting point, there is the sponsor, the project manager, and possibly the idea originator.
- The sponsor will start to invite people on the bus.
- The focus of the first stage (feasibility) is to validate the idea—that is, perform a feasibility study—and for this effort, the sponsor invites the feasibility team onboard.
- The driver is still the sponsor, with the support of the project manager.
- If the idea is feasible and management authorizes the project, the sponsor will hand over the wheel (i.e., control) to the project manager.
- The sponsor provides the project manager with the instructions on where to go and what to achieve.
- The sponsor and project manager also agree on the mandatory stops along this journey (the stage gates).
- The sponsor must stay on the bus, but the feasibility team gets off the bus, although some could continue in new roles.
- The project manager starts to take on other passengers—the project management team. This team will help manage the development of the project. The first activity is to plan the stage and decide on the other resources needed for development of the work (the execution team).
- Because this is likely to be a long trip, the project manager needs to stop every once in a while, for refueling, to conference with the sponsor, or even to ask for directions from people along the way (representing stakeholders who are not on the bus). During these stops, some people may get off the bus, and others may board.
- The project manager will stay in the driver's seat throughout the project development and most of the Delivery Phase—that is, for the implementation and close out.
- During implementation, the sponsor may form another team in the back of the bus to start learning about the project destination. This team also plans what to do on arrival. The sponsor will appoint a leader for this team; let us call him *operation representative*, or *operational readiness manager*.
- Once implementation is complete, the operation representative takes over driving the bus.

- At this time, most of the project management team members leave the bus, but the project manager (or another team lead) and a few others may assemble at the back of the bus to provide the necessary support when required.
- While the operation representative and his team are working on moving the bus to the final destination, the project manager and her team work on closing the project, discussing lessons learned, and gathering their maps, as a way of reflecting the potential changes that took place along the way.
- Once the bus is fully at the location and the end users are happy, the project manager will complete the project closure and depart the bus.
- *Trip accomplished!*

38.4 The Project Team

In earlier chapters, the project team topic came up. As we close this chapter, it would be useful to revisit this topic for a refresher on the membership of the project team.

The project team consists of the project manager, project management team, and technical/functional team. The extended team will also include the operational readiness personnel.

The project management team might be limited to the project manager only, in very small and simple projects. However, the project management team often consists of the project manager and various other resources who would help the project manager manage the project. These would include cost estimators and control personnel, planners and schedulers, site and project engineers, quality and procurement specialists, etc.

On the other hand, the technical team is not a consistent team, and its membership can change from one stage to another (as in the bus story). For some projects, outsourcing agencies could provide many of these team members. Furthermore, the types of positions vary greatly based on the project

---

A Personal Story

This personal story inspired the Bus Trip Analogy.

During a team-building session in the late 1990s that I was part of, the participants of the session had to give each other feedback consisting of a positive point and an improvement point. At that time, I was not too patient. It seemed that I always wanted to move fast on some decisions or actions, even when some team members might not be ready.

When it was my turn to receive feedback, a colleague, Wes Agnew, gave me his, "Mounir, you have to be careful, you might get to the bus fast, but remember not to leave until everyone else is on the bus."

I took that feedback to heart, and I recall it every time I'm in a situation in which I think we are ready to move but notice that not everyone is on the bus.

However, some people might take too long to get on the bus, so what do we do?

The way I work with any team is that I will be at the bus explaining and sharing to get people to join the trip. I will wait for others but only for a while. I tell my team, "Let me know if you are not ready to board and I will do my best to ease the transition and explain why we need to be on the bus." However, if some hesitate too much or cannot make up their minds, then we leave without them!

type. For example, in web design projects, the need is for web developers (programmers), graphic designers, copywriters, marketing specialists, online security experts, etc. Facilities projects require engineering and construction people in all disciplines.

It is vital to distinguish between the three types, because each group has its role on a project. However, on some small and simple projects, it is possible to have all three types combined as one team that will handle the project management, technical, and operational readiness work.

Chapter 39

All Other Topics and Layers

Other project management functions could have been addressed here, such as quality, safety, health, environment, procurement, and resources (other than human resources). However, the same concepts apply, because each of these functions has to be managed across the stages and per the three dimensions. The team will address them at the stage level, per the processes, and at the project level via the project life cycle.

Furthermore, topics such as Agile, program management, and portfolio management could have been explored further than just a mention. However, it was felt that the book must maintain focus on the Customizable and Adaptable Methodology for Managing Projects™ (CAMMP™) and the various dimensions and related components to provide the readers and practitioners enough guidance to start building or enhancing their organizational project management systems.

Section III, Part H
Practical Real-World Application

Chapter 40

Introduction to Applying CAMMP™

40.1 Introduction

Previous parts of this book addressed how to visualize the management of projects per the three dimensions, with a major focus on managing at two levels: the project level and the stage level.

In this last part, the focus continues on visualizing the application of CAMMP™ in the real world, but we will shift to address tailoring—customizing and adapting the model to different project types and classifications. We will also include brief simulations of projects from various domains. Furthermore, Chapter 42 will address the difference of applying CAMMP™ for service providers and project owners.

40.2 Specialization

This is an appropriate time to discuss a vital point related to the following hypothesis: for project management to grow and prosper beyond the current state, there is a vital need for specialization. Specialization in project management is essential for sustainable growth while pursuing efficient delivery of projects regardless of the domain.

> For project management to grow and prosper,
> there is a vital need for specialization!

One of the major disconnects today, which has become obvious in numerous online discussions, is that project management in commercial information technology—mostly software development—is entirely different from project management in the capital projects industry. It feels like professionals from these various domains do not even speak the same language anymore.

The battles for Agile versus Waterfall, change is good or bad, process groups versus project phases, common practice versus best practice, etc., etc., are just examples of the vast divide between these domains. In recent years, we see that even some professional associations are moving more and more toward commercial IT rather than *"most projects—most of the time."*

One way to achieve specialization is for some of the professional associations to consider either going beyond their generic focus or creating numerous organizations, each with its particular focus. Specialized certifications by industry might also help. Another way to encourage specialization is the main driver for this book and the Customizable and Adaptable Methodology for Managing Projects™. The CAMMP™ Model is an approach that would empower and enable organizations to start building tailored methods applicable to their specific project domains, types, and classifications.

Once again, organizations must consider more than one platform, guide, or approach to building a sustainable organizational project management system (OPMS) that incorporates various methods for the diversity of the projects they handle.

40.3 Project Samples

The following are some examples and references to help the reader visualize the application of CAMMP™. Unfortunately, it is not possible to provide details for these sample projects here.

A couple of follow-up books will cover case studies and full, end-to-end samples.

40.3.1 Sample 1

The author has published *Project Management Foundation* (Ajam 2014), the second edition of *The Inheritance* (Ajam 2010). The book is about a social entrepreneurship project to launch a not-for-profit organization that, in turn, is building a community center in a rural area. This sample simulates the whole project from idea to closure. The project combines a community initiative, with nonprofit, with a facility.

40.3.2 Sample 2

In another book, *Applying Project Management* (Ajam 2015), the author presents a case study of the completed project *writing a book.*

40.3.3 Sample 3

Chapter 44 includes a brief simulation for the project *writing and publishing a book.*

40.3.4 Sample 4

Chapter 44 will also discuss how to use CAMMP™ on an industrial project. The discussion provides a brief approach on how one can use the methodology on such projects.

40.4 Closing Comments

It is vital to build on what exists in the professional community and move toward specialization without forgetting the generic approach. A generic approach, such as those published by the various professional associations, is vital and an essential foundation. Organizations must build on these strong foundations by developing solutions that are custom-fit to their environments.

Professional associations can help, and guides such as the extensions to the *PMBOK® Guide* published by PMI are useful resources. However, the sad reality is that most practitioners do not know such guides exist. Marketing for certifications overshadows the existence of these resources.

Chapter 41

Managing per the Project's Class

41.1 Project Classification

In more than one place in this book, the terms *project classification* and *project class* were mentioned, and it is time to elaborate.

Projects are unique and vary from one sector to another.[1] They differ from one domain to another.[2] Even within the same organization, they may differ depending on the organizational functional unit leading the project and the types of projects it handles.[3] Furthermore, within the same kind of project (same type, domain, or sector) one may still have to differentiate between projects.

One way to distinguish projects and categorize them is a classification system. Such a system can be quite elaborate, with numerous factors and conditions. GAPPS (Global Alliance for Projects Performance Standards) offers a broad classification or complexity system. However, for the CAMMP™ model, there is a simplified approach that uses three classes:

- **Small-simple projects.** This class represents the smallest and simplest projects in an organization in relation to the other categories.
- **Medium-moderate projects.** This class is for larger, medium-size, and moderately complex projects in comparison to the previous category.
- **Large-complex projects.** This class would be for the largest and most complex projects within an organization. In some organizations, this may include mega projects, if not listed as an independent category.

---

[1] In this context, *sector* refers to government versus private organizations.
[2] In his context, *domain* refers to industrial projects versus technology projects or versus marketing projects.
[3] In this context, *type* refers to human resource projects versus financial projects.

It is vital to note that these classifications are directly linked to the organizational context. For example, a small project in one organization could be huge for another organization. Therefore, the classification has to be specific to the organization or type of work; otherwise, it is not valid.

41.2 How to Measure Size and Complexity

The foundation for the CAMMP™ classification system is two broad categories: size and complexity. The importance of the project can be another factor, but that is part of complexity.

41.2.1 Size

Size is usually related to cost, duration, and number of people involved, which is highly dependent on the organizational context, as mentioned already.

Each organization knows the type of projects it handles and the range of their (a) costs, (b) durations, and (c) people involved. Once the organization knows the range of projects per these factors, it can split them into three categories: small, medium, and large. They could also assign them numbers, for example, 1 for the lowest rank, 2 for intermediate, and 3 for high.

For example, fewer than 20 people[4] will have a score of 1; 21 to 100 people get a score of 2; and anything over 100 team members will be a 3. The cost factor will have a similar scoring system, as will the duration factor. Therefore, under the size category, there are three factors.

41.2.2 Complexity

Complexity can have more than one variable, such as (a) uniqueness of the project work, (b) importance, (c) number of departments involved, (d) new technology or not, and (e) organizational impact. In this case, one can consider projects as simple, moderately complex, or complex. In a similar fashion to size, an organization can also assign 1, 2, or 3 for each factor.

41.2.3 Summary

Therefore, this classification system integrates the size and complexity categories and may contain the eight proposed factors, more or less (see Table 41.1). These categories and factors apply to most projects. With the various factors and a numerical model, organizations can classify projects into these three project classes.

41.2.4 Scoring Model

Table 41.1 visualizes the classification system. One can change these factors, add or delete, or combine some of them. (N/A, not applicable, always works as well.)

---

[4] People in this context can be used for the number of people on the project team (PMT and technical) or the extended team.

Table 41.1 Project Classification Summary, Categories, and Factors

Category	Factor	Small-Simple (S-S)	Medium-Moderate (M-M)	Large-Complex (L-C)
Size	Cost			
	Duration			
	People			
Complexity	Uniqueness of the project work			
	Importance			
	Number of departments involved			
	New technology			
	Organizational impact			
Total Score		The sum of all points		

To complete this table,

1. The user can put a 1, 2, or 3 in the corresponding column (1 for S-S, 2 for M-M, and 3 for L-C, respectively).
2. With eight factors, the total score can range from 8 for the smallest, simplest project to 24 for the largest and most complex.
3. A total score of 8 to 13 leads to selecting the S-S classification, 14 to 19 will be M-M, and 20 to 24 for L-C projects.

Please note that these numbers and splits are rough and approximate rather than set in stone. Therefore, organizations can tailor this classification system to fit their environment. They could also use two classes instead of three, such as small to medium and medium to large.

41.2.5 Special Situations

There are situations in which one can have a project that is small in terms of size but quite complex technically, or a medium size project that is quite simple and routine. The numerical classification system can handle these situations. However, it is also possible for the team to decide how to classify these projects on a case-by-case basis. If the team wants to overrule the classification system—with justification and the agreement of the sponsor and the project management department—it would be possible.

41.3 Classification in Project Management Guides

Does the *PMBOK® Guide,* ISO 21500, ICB®, or PRiSM™ address a particular project class?
  Not really; remember these are generic guides and standards.
  Should these guides reflect all project's classes?

Yes and no. It is not practical (nor is it the goal of these guides) to address the details of various projects or how to classify them; this depends heavily on the organizational context. However, these guides can offer some suggestions.

We understand that the *PMBOK® Guide* sixth edition includes tailoring as a subject in the various knowledge areas. However, that is somewhat different than what we are talking about here. CAMMP™ is primarily about tailoring within the organizational project management system (OPMS), which is specific to establishing standardized methods and processes to use across the organization for the various project types and classes. If that is done properly, then the project management team can do further tweaking or minor tailoring, if necessary, on a project-by-project basis.

Please refer to Chapter 43, in particular, Sections 43.8 (page 281) and 43.10 (page 282).

On the other hand, tailoring in the *PMBOK® Guide* is limited to the work that the project manager will be doing to manage a specific project.

Consequently, organizations pursuing change initiatives through projects and seeking benefits through projects must consider the project classifications and establish an OPMS that reflects the various projects' classifications.

41.4 How to Manage per a Given Classification

41.4.1 Overview

Here is the magic of CAMMP™.

The suggested model offered next takes into consideration the integration of the first two dimensions of CAMMP™. The approach depends on the question of how to integrate and use the processes within the project life cycle.

The third dimension is also present and would apply on all of the approaches listed below.

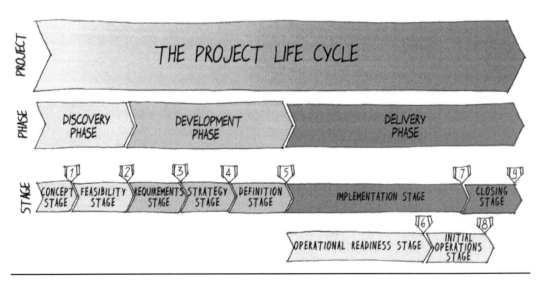

Figure 41.1 The CAMMP™ model for small-simple projects.

41.4.2 Small-Simple Projects

It goes without saying that small-simple projects can be handled in a simplified and practical way, whereas medium and large projects require significant effort.

For small-simple projects, the CAMMP™ approach is to use the first dimension, a tailored project life cycle model, as the starting point. Figure 41.1 presents the standard model, before tailoring. The project life cycle would include all of the appropriate stage deliverables and gates. This is the first step.

The next step is to consider all of the processes per Chapter 17 and use them where applicable. To use these processes and their process steps, the organization can spread them across the project life cycle, where appropriate. For example:

- Project authorization will include the Authorize process and its steps.
- The Strategy Stage will include the Plan Management process.
- The Definition Stage will use the Plan Details process.
- For project control and change management, the team can use the model described in Chapters 35 and 36 along with the Control process from Chapter 17.
- The same concepts apply to risk management, estimating, people aspects, procurement, quality, and all others.

41.4.3 Medium-Moderate Projects

As a project becomes larger (mid-size) and with moderate complexity, the approaches presented for small-simple projects might not be enough and may offer unsatisfactory results. For these projects, the recommendation is to integrate the set of processes into the project life cycle, *at the phase level*. Remember, *the CAMMP™ model separates the stages from the phases*.

Figure 41.2 presents this concept. Notice that the processes repeat in every phase.

This figure represents the CAMMP™ project life cycle with the three phases. However, organizations can use this concept with any other project life cycle, if they prefer to use an alternative.

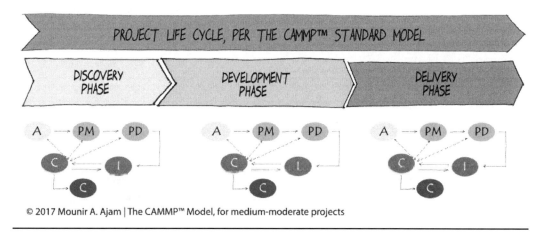

Figure 41.2 The CAMMP™ model for medium-moderate projects.

266 Project Management beyond Waterfall and Agile

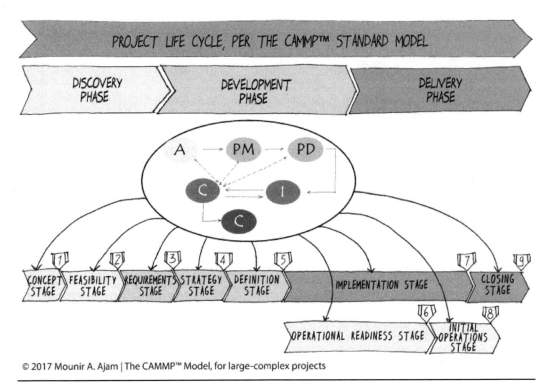

Figure 41.3 The CAMMP™ model for large-complex projects.

The key is to have the processes repeat as shown. Keep in mind—the third dimension applies on top of the first two dimensions.

41.4.4 Large-Complex Projects

For large and complex projects, the situation is even more elaborate. These projects can be substantial, take a long time, cost lots of money, and have numerous individuals and stakeholders involved.[5] In these projects, each stage could be months in durations and cost millions. Therefore, the recommendation is that for these projects, repeat the processes in each stage. This concept is the subject of Figure 41.3.

41.4.5 Blended Approach

Applying the above concept following a blended approach is also possible and might be the most likely scenario.

For example, the team may choose to use the processes at the phase level for the Discovery Phase, not for each of its stages. They can do the same for the Development Phase—applying them only at the phase level, not the stage level. This approach is acceptable, assuming these

---

[5] Can be in the hundreds if not thousands.

phases are not highly complex or require significant resources. However, for the Delivery Phase, which can be quite resource intensive, the team can apply the process groups for each stage.

41.5 Closing Comments

Remember two fundamental principles:

1. Project management must follow a disciplined approach, and
2. Project management must be a *facilitating* process and not a *controlling* process.

These principles mean that organizations must have discipline through a proper system but allow flexibility, because not all projects can be locked in one category or another. Furthermore, if project management becomes bureaucratic or requires too much control, the team will become slaves to project management instead of using project management principles and skills to serve their organizations.

What CAMMP™ offers is a system, a standardized yet adaptive approach, which is an essential component of the OPMS. However, a standardized approach is not set in stone, and one-size-fits-all is not an option. The team can overrule the system, but only with justification and agreement with the sponsor and project management department, which should own the OPMS. In other words, delegation, empowerment, and management by objectives are all necessary concepts for managing projects, but not at the price of chaos and a loose system. Loose systems will encourage some team members to take shortcuts and bypass critical steps that would be vital for the successful delivery of projects.

If organizations do not want to follow these principles and do not want to enforce the use of the OPMS, they might as well not build one.

Chapter 42

Life Cycles 360°

42.1 Introduction

The two words "life cycle" are used together so often, and in different contexts, that confusion is a normal result. Because the Customizable and Adaptable Methodology for Managing Projects™ (CAMMP™) is based on the *project* life cycle, we need to explain what these words mean in specific contexts.

Life cycles include product life cycle, project life cycle, project management life cycle, and life-cycle costs.

42.2 Product Life Cycle

42.2.1 Defining "Product"

Before explaining the term *product life cycle,* let's examine the word *product*. Let us start with a question and a scenario, which we will use throughout this chapter.

The scenario is a project to build a water bottling plant.

What is the product?

The answer might be "water bottles."

However, this is a trick question and may lead people to seek clarification.

A proper response would be: "Not sure, is this about the product of the project or the product of the plant? If it is the product of the *plant,* then it is bottled water; if it is the product of the *project,* then it is the plant itself—the physical facility."

That is a great answer!

In this regard, the *product of the project* is the plant facility. This facility, once operational, will deliver the *consumer product,* which will be the bottled water.

So with this understanding of the word *product,* what is the product life cycle?

42.2.2 Product Life Cycle

There are plenty of publications on this topic, and we encourage readers to seek them out for in-depth knowledge. The product life cycle extends across the whole life of the product, regardless of what that product is. In this context, the product is the output that the project delivers to the organization. It can be a software application, a book, a new organizational system, or, as in this scenario, a water bottling plant.

The product life cycle is a time span from the moment the idea for the product is identified until the dissolution of the product in one way or another—in other words, the end of its *useful life*. This useful life could be a few years for a software application or a car model, or decades for physical facilities such as buildings and industrial plants.

Throughout the duration of useful life, there are shorter periods called *phases of the product life cycle*. These are product life phases and not project or CAMMP™ phases.

"OK, English now please!"

Let's return to the water bottling plant scenario as the example to clarify this point.

1. An investor has an idea to build a water bottling plant. A given geographical area does not have one, and the investor believes there is a market for it. This is the idea for the product, and it is, coincidentally, the idea for the project. The starting point of the product life cycle and the project (investment opportunity) is here.
2. The investor team would examine the idea (investment opportunity) through a feasibility study. If the project is feasible, the investor will likely go ahead with the project (pursue the opportunity) and progress through the project development and delivery phases, per a model such as CAMMP™. At the completion of the project, the plant is built and then handed over to operations, which can start normal operations.

   So far, the *product* life cycle and *project* life cycle are essentially overlapping, but they are not the same. With final project acceptance, the project team starts closing the project, ending the project life cycle, while the product life continues.

   In the context of the product life cycle, the project, in its entirety, is only the first phase, the Acquisition Phase, of the product life cycle.
3. At the end of the product life cycle's Acquisition Phase (the project), operations will have to operate and maintain the facility for its useful life.

Based on the above, one can define, simplistically, that the product life cycle has two phases: Acquisition and Operation. The transition point between them is the product acceptance, although operations involvement starts pre-handover. A project such as the construction of a manufacturing plant could last two to three years. However, operations can continue for 20 or 30 years or more.

42.2.3 Product Life Cycle: Textbook Perspective

In the previous paragraph, the product life cycle was split into two phases: Acquisition and Operation. This split is an oversimplification. Most literature on product life cycle defines it as a cycle of five phases, with the following common names applied to these phases:

1. **Development.** From idea/concept, acquisition, and even initial operations, this is what was called the Acquisition Phase in 42.2.2.
2. **Growth.** Increased sales, growing demand for the consumer product.
3. **Maturity.** Good production system, product well known in the market.
4. **Decline.** Declining demand, building/plant is too old, it costs more to operate and maintain than would be justified, the product is losing its value—which leads to the final phase.
5. **Withdrawal.** The organization takes the product of the market.

Activities within and drivers for each of these phases will differ depending on the nature of the product. In general, an organization could deliver numerous projects during the product life cycle and its phases. For example, during the growth phase, there could be projects to expand the facilities and increase production or add features to a software or business process. There could be maintenance projects during any of the phases. At the end of useful life, there will be a project to withdraw from the market or, if a facility, to dismantle or shut down.

42.3 Project Life Cycle

The project life cycle has been adequately covered previously. Remember, CAMMP™ adopts the view of the project life cycle, end-to-end, from the perspective of the *idea owner* or *project developer*—the project owner.

Is this always the case?

No, this is only one perspective.

"I am a service provider, why are you only viewing this from the standpoint of the project owner? How does my point of view fit in?"

Please don't worry, we're visiting you soon. Furthermore, who said that the above does not have you in mind? If you consider yourself a "provider" of goods or services, we ask you to have two hats handy:

- **The first hat.** Regardless of who you are and what your organization does, you have your internal projects. In those cases, *you* are the project owner. Everything in this book applies to you, do not you agree?
- **The second hat.** This second hat is the one you wear when you're working to deliver a service to a client—the project owner. We cover this perspective next.

42.4 Service Provider Perspective

42.4.1 Overview

In response to the previous section, a colleague offered a challenging question (refer to the Reviewer Question), and here is the answer.

> **Reviewer Question**
>
> "Consider the water bottling plant scenario. The owner outsources a service; why can't we consider that the contractor or service provider also applies the same project life cycle model? This way the model applies equally well to all parties."
>
> The ultimate objective is to help project management practitioners enhance their own, and their organization's, project performance.

Based on the intent of CAMMP™, which includes the ability to customize and adapt the model, service providers *can* and *should* follow this systematic approach because it applies to them, but not in the standard format—some tailoring is necessary. Let us discuss this in detail.

Mr. or Ms. Provider, keep in mind that the following is from your side of the fence when you wear the *service provider* hat!

42.4.2 The Sample Project

The following sections and discussion are for a design/build (D/B) contract for the water bottling plant scenario. The D/B contract is specific to the project owner's Implementation Stage, but it is the service provider's project, end to end.

42.4.3 How Many Stages?

From the perspective of the service provider, is this a one-stage project?

Is it two: design stage and build stage?

Are there more than two stages?

In other words, how many stages does the service provider have?

It must be clear by now that, from the project owner's perspective, this is only one stage out of nine per the standard model. However, from the service provider's perspective, it is a project, with its own project life cycle, like all projects.

How could this be, a project within a Project Stage?

The explanation below demonstrates this from the service provider's perspective. Let's view the project life cycle of the service provider, stage by stage.

42.4.4 Discovery Phase

The CAMMP™ Discovery Phase sequence is project brief, SG1, feasibility, SG2, and finally project authorization document (PAD).

Let us apply this to the example at hand.

As a provider of a service, the idea statement—the project brief—is simply an external request for proposal (RFP) from the client (project owner); in other words, the idea source is external to the provider organization. To confirm, the RFP is the official starting point.

Per CAMMP™, the idea must pass through Stage Gate 1 (SG1): in this case, SG1 of the provider's organization. Yes, providers should have a stage gate, because they can reject work. At this provider gate, the provider management decides whether to accept or decline this RFP, for whatever business reasons they might have.

If provider management accepts the RFP at SG1, then the provider should do its feasibility study. Otherwise, the provider might agree to a project beyond what it is capable of handling. This feasibility study has a different focus from the owner's study. This study can be quite simplified for routine and repetitive projects.

Once the provider's feasibility study is complete, it's time for the second gate: the final decision on how the provider will respond to the RFP. If they decide to propose, they will notify the client of their intent to submit a proposal. If no, they will decline the work and the work stops.

All is smooth so far and matches the logic of CAMMP™. Now the trick question: after Stage Gate 2 (SG2), can the service provider issue its own project authorization document?

The answer is that for the overall project—the project by the project owner—the answer is no. There is one PAD per project, written from the perspective of a given organization. In other words, the project owner has its PAD, which is independent of any service provider.

"OK, we hear you, but I am not the project owner, I am the service provider; don't we have an authorization document?"

Of course! At the end of SG2, and if the service provider's management decides to respond to the RFP, the service provider's executive management must issue a PAD for its organization.

"But—how can we authorize the project when we do not have a contract yet?"

As far as the service provider is concerned, the internal PAD is necessary to authorize the commercial and technical teams to start working on a response to the RFP. In other words, the project is being launched by the service provider organization and there is no need for a contract, because this work is "at own risk."

Other possible names for this stage are pre-sales, prospect stage, or anything else that fits the type of project. Regardless of what you call it, it is in total alignment with the Discovery Phase as prescribed by CAMMP™. The only difference is that the idea is from an external source, which is a client's RFP.

42.4.5 Requirements Stage

The Requirements Stage calls for defining the project requirements then moving to SG3.

If we relate this to the example, what are the steps?

In this case, the service provider has an essential part of the big picture—the project from the perspective of the client. The provider has to understand the output of the project—what the project owner wants: to learn about needs and expectations, define the requirements, and ensure alignment with the client (project owner). All of this leads to the provider's developing its internal project requirements document (PRD) to enable its team to work on the project.

42.4.6 Strategy Stage

The Strategy Stage calls for developing a project management plan (PMP) per the CAMMP™ guidelines, which will lead to the provider's SG4. Again, this is internal to the provider.

42.4.7 Definition Stage

The Definition Stage calls for developing the project detailed plan (PDP), then approval at SG5. Remember, SG5 is also the point of funding approval per the standard CAMMP™ model.

From the service provider's perspective, the Definition Stage contains the necessary detailed planning activities leading up to submitting the proposal. It may be the case that for large or complex projects, the service provider has to develop a "solution" to include in the proposal. To do this, the service provider will have to go through this stage and develop the detailed scope, schedule, cost estimate, resource requirements, risk management, subcontracting strategy, etc.

Based on all of the above, the service provider will determine its costs and decide on the offer price for the proposal as a final step before submission.

Once the proposal is in the hands of the client, there is the evaluation and possibly a negotiation period. If the customer awards the contract, the provider will have the opportunity to move into implementation. If not, the opportunity is lost. In this case, the provider will move to project closure without passing through the Implementation Stage.

42.4.8 Implementation Stage

Generically, once the client awards the design/build[1] contract to the service provider, the provider will proceed and implement the scope of the contract, whatever that scope is; in this case, it is the full engineering design and construction of the water bottling facility. In this period, the owner and provider Implementation Stages are joint, but it is likely that most of the work is done by the service provider; the project owner might have other activities, in addition to oversight.

42.4.9 Operational Readiness Stage

The Operational Readiness Stage is likely to be led by the project owner directly, with support from the service provider if necessary. In other words, it is probable that the service provider will not have any work here and will not have an Operational Readiness Stage in its own project life cycle.

42.4.10 Close Stage

The service provider will perform its Close Stage, independent of the project owner's project closure. A close-out report will include reconciliation, lessons learned, and update of the organizational records.

42.5 General Observations

42.5.1 CAMMP™ Works for Both Perspectives

The reader may have noticed that the service provider's perspective of its project life cycle is quite similar to the project owner's, and in some cases they move in parallel. This situation is a demonstration that CAMMP™ works across different perspectives.

Furthermore, in the design/build contract case, the Implementation Stage for both the owner and the provider are essentially the same, but inevitably differ in some respects such that each entity will look out for its own interests.

---

[1] Remember, the example is about the design/build contract for the Implementation Stage of the water bottling plant.

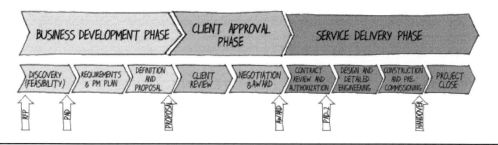

Figure 42.1 Project life cycle, service provider perspective.

42.5.2 Two Projects or Two Phases?

Another observation one might make is that from the service provider's perspective, "the whole thing," from RFP to closure, can be viewed as one project with two or three phases. In other situations, the provider might split the whole thing into two projects: the business development project and the service delivery project.

Figure 42.1 shows the first option, one project for the whole thing with three phases. In this perspective, the service provider considers business development and service delivery as two phases within this project, separated by the client approval phase. In turn, each of these phases has its stages and gates. Figure 42.1 is just an overview, and it does not show the stage gates, the process groups, and all of the work within the stages.

Also, notice that there is a project authorization at the end of the Discovery Phase, but there is also another authorization, PAD-2, in the service delivery phase and after the contract review. This is necessary for two reasons: (a) the actual contract could have variations from the proposal, and (b) this work is now under contract, whereas the first PAD was work-at-risk internal to the provider.

42.6 Closing Comments

In closing this chapter, suffice it to say that *CAMMP™ is a methodology for managing projects, regardless of who does what.* The project owner, service providers, even vendors and subcontractors can use this systematic and practical approach. If an organization delivers projects, small or large, internal or for external clients, CAMMP™ is an approach that can be tailored to fit the organizational needs.

Tailoring is the subject of the next chapter.

Chapter 43

Customizing and Adapting CAMMP™

43.1 Need to Tailor CAMMP™

The need to tailor CAMMP™ has been explained in earlier chapters. CAMMP™ uses the terms *customizable* and *adaptable* as a two-step tailoring approach.

To customize the standard model is to modify it to fit specific domains, which is necessary for project success. The model must be customized to an industry and a given organization to fit its project's specific needs and requirements.

On the other hand, *adaptability* means to adapt the model for different classes of projects within the same organization and its various functions.

Although these are explained as two separate concepts, they are quite interlinked. Therefore, one alternative is to combine the two concepts into one change initiative in which merging them makes sense.

43.2 Example to Differentiate

Many organizations in the capital projects industry follow a project life cycle approach for managing projects. These organizations—at least some of them—*have customized methods for their capital projects.* The project life cycle across the various companies in this industry would be similar, almost identical, except that the names of the stages could be different. In addition to customization, they *adapt their approach and have a method for large projects and another one for small projects;* they typically use dollar value as a differentiator.

For example, one company—say, a petroleum company—may have a seven-gate model for major projects and a three-gate model for small projects. In both cases, these are capital (engineering and construction) projects, but due to their differences in size and complexity, they need to customize and adapt the model. In other words, the project domain requires customizing the model to industrial, capital projects, and the project size (and, indirectly, complexity) is the main classification that drives adaptation.

One cannot presume to judge if three gates are enough or if seven gates would be appropriate. It is a function of the organization's preference and culture, as long as they remember that project management must follow a *disciplined approach*! To summarize these points, stage gates are a must; the number of stage gates is per organizational preference.

Furthermore, the same organization considered their information technology, marketing, finance, and other departments' needs for a methodology. Given that the projects from the other functions were quite different, it did not make sense simply to copy the capital projects model. They had to customize and adapt a model for their functions.

43.3 Why Customize and Adapt the Model

Let us say that an organization has accepted project management as a strategic capability and function within the organization and requires the use of the principles of project management to manage the organization's various projects, regardless of the functional unit.

For such an organization, it is likely that any change initiative, new idea, new human resource system implementation, or development of a new software tool, are all treated as projects and must be managed in accordance with an established methodology. Then should every human resource project, software project, and even construction project follow the same standardized project life cycle and method?

No!

What this situation requires is not just a common approach for control and governance, but also customized and adapted approaches appropriate to the types of project and their classification.

A refresher:

1. Project management has proven its value across industries, organizations, and functions.
2. Project management must be a *facilitating* process, not a *bureaucratic* process.
3. Project management must be flexible, yet not loose and without any control.
4. Project management is about discipline: establish a system/process and follow it.
5. *One size fits all is not* for project management.

Then how to honor these important factors and cater for different divisions within an organization? By adapting the model with the principles listed above as guidance. Therefore, organizations can:

1. Select a methodology.
2. If an organization does not choose the CAMMP™ model, management must ensure that the methodology they choose incorporates CAMMP™ essential principles.
3. Then customize the methodology to requirements.
4. Finally, adapt it to the various internal divisions.
5. Even within the same division, one may have to adapt it to different project classifications.
6. As a result, organizations will likely finish with a few "standard" models, one for each project type and class.
7. Once established, follow the tailored models.

43.4 What Is Customizable and Adaptable?

Following are the most frequent types of customization:

1. It is an organization's choice to have both phases and stages, as shown in CAMMP™; the alternative is to use either phases *or* stages.
2. The name of the stages and phases are a function of the industry, typically, and organizations can change them.
3. The number of stages can be reduced only by merging stages, not eliminating them.
4. For certain projects, there is nothing to operate post completion, and the Operational Readiness Stage and Initial Operations Stage will not be applicable.
5. A reduction in the number of stage gates is possible, but only by merging the control aspects (the function/purpose) of an eliminated stage gate with the stage gate that follows.
6. The number of estimates can vary from the model basis. However, it would be better, from a governance and control perspective, not to use fewer than two estimates.
7. On some projects, it might be necessary to split a stage into more stages or substages.
8. Some organizations may treat some of the stage gates as informal gates. This informal practice may be acceptable, but be cautious not to make the stage gates too loose or too easy, or they will lose their effectiveness.
9. Instead of eliminating stages, one can increase their overlaps to fast track a project. This fast-track approach may be necessary for projects in which expedited delivery is a critical success factor.
10. There may be other customization opportunities, such as merging operational readiness with implementation or starting the operational readiness work before implementation, among other possibilities.

43.5 Customizing to a Project Domain (Industry)

Projects from the same domain, industry, or application area share similar project life cycles. The stages to publishing a book would be similar for most publishers. The stages of capital investment projects would also be similar. The same would apply to web development, marketing, information technology, and other projects. Therefore, there could be customized models for given industries.

In line with the above, an organization can search for models from similar domains before customizing CAMMP™ to its situation. The current challenge with this scenario is finding enough published information from the various domains for organizations to use as a starting point.

Another key challenge here is the fact that customization is not only about redrawing the standard model, which is easy. Customization includes modifying all of its components, such as processes, forms, templates, criteria for stage gates, criteria for project success, and numerous other elements of the organizational system.

43.6 Common Concepts

It is necessary to stress the key concepts to customize and adapt the model, which are common across domains and project classes.

1. Every project must have an idea that is documented, reviewed, and approved, with a validated alignment to the organization's strategic objectives.
2. Every project must have a feasibility study to determine whether the project is commercially valid and whether the organization can deliver it successfully.
3. It is recommended that organizations assign a project manager, or involve the project management function, during the first phase and conceptual/feasibility studies.
4. Once an organization authorizes a project, it must do so through a formal document.
5. Before proceeding, every project must have the project expectations explored, requirements defined, and both agreed to by the various stakeholders.
6. There is a need for a project management plan (PMP) and a project detailed plan (PDP). In some cases, these two can be merged, but it would be best to keep them separated, even for small and simple projects.

The list can include more items, but we think it is already clear that this list represents the steps of the model—the various stages.

43.7 How to Customize

An organizational change initiative is the recommended approach to customizing the model for an organization. When an organization decides to implement an organizational project management system (OPMS), including a methodology and related processes, this is the time to initiate the project to adopt CAMMP™ (or another model) and customize it to the organizational needs and requirements.

43.7.1 Similar Projects

If an organization's projects are all similar in nature, whether construction, engineering, marketing, training, business development, publishing, events, or any other type, then the leading practice is to modify the standard model to fit the types of projects that the organization works on.

Another leading practice is to adopt and customize the model and to use it on one or two pilot projects to capture learning, make modifications, and then roll the enhanced model out across the organization. This way, the chance of success is improved.

43.7.2 Different Projects

If the organization's projects are not similar, and the organization wants to deploy the model across various functions, the approach needs to be different, and a two-step process is used.

- **Step 1** is agreement on the key elements that are necessary across the whole organization—the common elements. For example, use phases and stages or one of the two, need a stage gate at closure or not, is there a common point for project funding, and anything else that would apply for all projects within the given organization.
- **Step 2** takes place at the functional unit level. Each functional unit continues the customization to fits its project types. Alternatively, this step could be part of adapting the model to a given function.

43.7.3 Who Will Customize

If the organization has a project management department, a project management office, or something similar, this would be the appropriate unit to lead this change initiative to implement a project management organizational system, including the methodology.

If such a unit does not exist, the organization may hire a consultancy to do so, but we think this should be the last resort. It is better that the organization establishes its project management unit and then has this new function take the lead. However, if necessary, consultants could be engaged to support and guide, even lead.

In either case, this functional unit will own the methodology and handle future enhancements and updates. In small organizations, this project management unit might be a single person functioning as an internal consultant.

43.7.4 A Reviewer Comment

One of our reviewers provided this comment on the name of the model, especially the concept of customization. His comment was, "Is not the whole concept of a project management methodology for it to be customized to the organizational environment?"

Absolutely!

It was stated before that one reason it is not common to hear or read about published methodologies is the fact that methodologies should be custom-fit to organizational needs and requirements. However, the challenge is that a significant percentage of project management practitioners do not understand this point. They even look at the *PMBOK® Guide* as a methodology, even though it is not. Some practitioners keep asking for forms, templates, even written samples that "must apply to my area." Because such practitioners do not fully understand this concept, they overlook the need to customize standards to their own or their client's environment, whether the standard is a methodology or a framework.

43.8 Project Manager Authority

When the model advocates tailoring, does it mean that every project manager customizes the model as he or she likes?

No!

Giving this level of power to the project manager is one of the potential pitfalls, which we will emphasize in the Potential Pitfalls section below.

A blog post once addressed this point indirectly. The question posted in the blog article was: "Who should decide on (as in design) the project life cycle for projects?" Most of the answers were that the project manager is the person to design and decide. There is no question of the value of the project manager in this process, but what most answers presented is the gap in the current state of practice in which organizations do not have standardized processes nor do they have well defined methods or life cycles for their projects.

Yes, projects are unique, but each organization often has the same types of projects, be it marketing, training, engineering, facilities, publishing, etc. Is it logical to recreate the wheel with every project or to establish a standard life cycle, giving the project manager and team time to focus on what is unique in the project? There's time for innovation and creativity in the solution.

43.9 A Story Reflecting Current Reality

In the midst of editing this work, there was a post on a social media group in which a project management consultant asked the following: "Where can I find a complete case study for PMP training . . . my students keep asking me to provide actual case study . . . Thanks."

How can anyone provide a "complete case study"?

A case study of what, specifically?

- To study for the PMP® certification, or a case study for a project?
- If a case study for a project, then from which industry?
- From a service provider's or project owner's perspective?
- An internal business project or a facility project?
- Is the project to deliver a product, a service, or some other result?

The intent is not to complicate things or be difficult, but such a question comes from a project management mindset of *one size fits all,* which is clearly not a proper choice. Questions like this were one of the drivers behind the CAMMP™ model and publishing program!

43.10 Potential Pitfalls

Like any methodology, framework, or system, there are potential pitfalls. This section addresses the potential pitfalls unique to CAMMP™, which could occur if the model is not understood or used properly. The words *disciplined approach* have been mentioned many times in this book, and this is likely to be the last time. As a refresher, what does this concept mean?

- A disciplined approach means that an organization decides on a methodology that applies to its industry and circumstances and follows it.
- The customizable and adaptable approach is there for organizations to consider. They can implement the standard model to the various situations and classes of projects in their organizations after systematically establishing alternative models.
- It also means to establish the above in the OPMS; in other words, to institutionalize it.

We are still reading the sixth edition of the *PMBOK® Guide,* which now includes tailoring across the guide and the standard. However, the guide emphasis and focus is on tailoring done by the project manager. This is an area which we highlight as a potential pitfall—giving more power than necessary to the PM is a recipe for possibly challenged and even failed projects. As we present, the PM has an active role to play, but after using already tailored models from the organizational system. If a project is truly unique and no organizational model exists to fit its management, then the PM can either tweak an existing model or tailor a new model. If tailoring a new model is necessary, then this new model should become part of the OPMS.

With the above established, the following are some of the potential pitfalls, although this is not an exclusive list.

1. **Too much flexibility.** Because the model advocates flexibility through customization and adaptability, users may be tempted to take these concepts too far and start to modify the model for their convenience, instead of what is appropriate.

2. **Ad hoc deviation.** Project managers must not deviate from the established methods from one project to another just because of personal preference. If there are good reasons to deviate, organizations should have a process in place to do so, and not leave it up to individual project managers to decide. This flexibility is one of these things where the project manager *cannot* decide but *can* recommend.
3. **Skipping important functions.** Some may misunderstand the concept of customizable and adaptable and think that it is okay to cut stages or stage gates.

   As mentioned earlier, it is okay to reduce the number of stage gates by combining them, but notice the word *combine* and not *skip*. In other words, defining requirements for a project is a necessary step. It does not matter whether there is a gate at the end of this or that deliverable, or if one combines that gate with, say, the PMP stage gate, as long as the team recognizes that by not having a stage gate at the end of the Requirements Stage, they risk misunderstanding the expectations and needs, which will result in a PMP that is not accurate.
4. **Loose stage gates.** Often organizations adopt the stage-gates concept but allow the stage gates to be quite loose or permit project managers to approve their stage gates.

   For example, let's say the team has developed the PDP, which leads to the highly critical final approval and funding gate. Here, the project manager can say that the PDP is complete, but is it truly complete? Is it done per the organizational criteria for a proper PDP? Specifically, is the PDP complete per the documented requirements, such as schedule, cost, and other key components, so that the result is both a higher level of confidence in the PDP and reduced threats in the next stage?

   Based on the above, stage gates must be robust enough and have clear and unambiguous requirements on what it takes to go through them! Some organizations have a "gatekeeper" role to avoid such situations. The gatekeeper is an organizational role to ensure that stage-gate criteria are met before this deliverable is brought in front of the stage-gate review committee.
5. **Not having an owner for the methodology.** It is quite important for any organization, whether it has a project/program management office or not, to have an owner for its processes. The process owner is typically an organizational unit that is responsible for the project management methodology and its applications in the organization. This owner also documents any requirements for changing the model for different applications that the organization might not have considered. It is also important for this owner to maintain a database of lessons learned from projects, as they relate to the model.

It is likely that there are other potential pitfalls related to the model or project management in general. Organizations must watch for these pitfalls if they are to advance project performance and, consequently, organizational performance.

43.11 Closing Comments

One standard model does not work, unless all of the organization's projects are the same.

Reinventing the wheel with every new project is not a good practice either.

Therefore, it is necessary for organizations to establish an organizational project management system (OPMS), which includes various methods, one for each type or class of projects that they handle. Numerous studies by leading organizations, including the Thought Leadership Series (Project Management Institute 2013b) published by PMI in partnership with The Economist Intelligence Unit, Forrester Consulting, and Boston Consulting Group, clearly shows the necessity for having standardized project management systems.

Chapter 44

Simulating CAMMP™ for Various Projects

44.1 Introduction

Before closing this book, this chapter will take two possible projects and briefly expand on the description to show the use of CAMMP™ on projects that are from different domains, types, sizes, and complexity.

The projects are a capital investment project and a writing and publishing a book project. Please note that the explanation below is not a full sample project; it only includes some of the concepts of CAMMP™ to demonstrate how one can apply the model on realistic projects.

44.2 Capital Investment Project

A capital investment project is typically a project, which would include the construction of a facility, such as a factory, oil refinery, power plant, hospital, or similar projects.

44.2.1 The Project Life Cycle

Figure 44.1 illustrates the project life cycle for a capital project. Notice that it is quite akin to the standard model with some minor differences, mostly terminology related.[1]

What are these differences and items to highlight?

---

[1] The three estimates (classes 1–3) are also shown in this graphic in the middle of the relevant stages. In reality, these estimates should be closer to the right side, because they would be developed closer to the end of the stage, but that space is used for the stage deliverable.

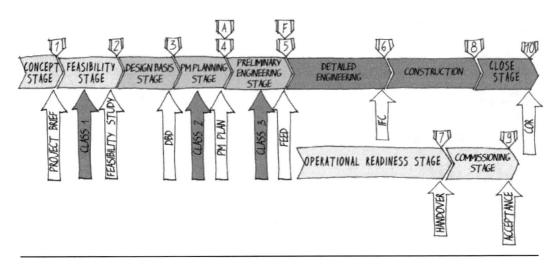

Figure 44.1 Sample project life cycle for a capital project.

44.2.2 Discovery Phase

There no major differences in the Discovery Phase.

The class 1 estimate is the first estimate in the project life cycle. The team uses this estimate during the feasibility study, for economic analysis, and it would be a key deliverable before Stage Gate 2 (SG2) for the initial decision. In these projects, a class 1 estimate is necessary!

44.2.3 Development Phase – Part 1

This part covers the Requirements and Strategy Stages, and the key highlights are:

1. This model uses the term *design basis* (DB) instead of Requirements Stage and *design basis document* (DBD) instead of project requirements document (PRD). Other terms used by organizations include *design basis memorandum, design basis scoping paper,* and *master plan,* among others.
2. As soon as the DB work progresses, the project management personnel would start working on delivery and management strategies, leading to the final project management plan (PMP) after the completion of the design basis document.
3. The DB work typically focuses on the major equipment for the project, the heat or energy balance, tentative plot plan (showing locations of major hardware), equipment layout, and other technical aspects.
4. The class 2 estimate, before Stage Gate 4 (SG4), is likely to be ±15%, ±25%, or similar ranges. Organizations use the class 2 estimate to determine the amount of advance funding required, which they would approve at SG4 if the project is a go.

44.2.4 Development Phase – Part 2

This second part is the Definition Stage. The major highlights are:

1. The model uses the term *Preliminary Eiingineering Stage,* and the stage deliverable is likely called *front-end engineering design* (FEED), *front-end engineering package* (FEEP), or *process design specifications* (PDS).
2. The main work in this stage includes many activities, but the primary one is the industrial design—design of all the main equipment, major piping, process control, instrumentation, power requirements, and other main components; but there is no detailed engineering design yet. Detailed engineering is part of the Implementation Stage.
3. The final estimate (class 3) would be progressing in parallel but cannot be completed until the end of preliminary design. This estimate is usually ±5% or ±10%. It would also serve as the basis for final funding.

44.2.5 Implementation Stage

Moving on to the Implementation Stage:

1. Implementation (or execution) includes the detailed engineering, procurement, and construction work.
2. The figure shows two stages for implementation: one for detailed engineering and one for construction. The stages are separated by a stage gate, SG6 in this case.
3. The owner organization could award these services as one contract, the engineering-procurement-construction (EPC) contract, or another common type of contract, the design/build (D/B) contract. EPC is a common term in the industrial/petroleum industry, whereas D/B is common in real estate or buildings projects.
4. Alternatively, the client could split engineering (design) from the construction (build) contracts. It is also possible to have multiple construction contracts or subcontracts. What action the project owner takes depends on the complexity and size of the project, along with the organizational policies.
5. With one contract for EPC or D/B, SG6 (refer to Figure 44.1) would become less important than if the client had separate contracts for engineering and construction, in which case SG6 would be crucial, because it allows the transition of the project from one provider to another, and there are contractual implications to consider at this interface point.
6. During construction, the pre-commissioning activities are taking place. The contractor would perform these activities, but they would be witnessed/accepted by the client's project management team and operation representatives. Pre-commissioning activities include testing the control loops, the electrical circuits, the piping, and other systems.

44.2.6 Operational Readiness Stage

For these types of projects, there will be an Operational Readiness Stage, and the key highlights are:

1. During construction, and even earlier, *planning for startup*—the term used for starting the plant operation—would commence and would include writing operating procedures, training the operators, control simulations, among numerous other activities.
2. It is okay to start performing pre-commissioning activities, mostly to test that all systems are connected and will function properly.
3. Once operation personnel are ready to take over the custody of the facility, operations (the client) would issue mechanical completion certificates to the contractor(s).

44.2.7 Initial Operations Stage

The team can combine the Initial Operations Stage with the previous stage or keep it independent. In this case, it a separate stage.

1. With the custody transfer at handover and mechanical completion, operation's personnel start final commissioning, leading to the official start-up of the facility.
2. Final acceptance could be a year later, usually after conducting a final acceptance test to demonstrate that the plant is producing at the design capacity with no issues.

44.2.8 Close Stage

In parallel to initial operations, the project management team would commence project close to complete it after final acceptance.

44.2.9 Project Success

Project success is not a stage, but it is included here as a separate topic because of its importance. How does one measure success on such a project? By using the four dimensions described below.

Product Delivery Success

The output of the Preliminary Enigineering Stage (Definition Stage in the standard model) is a set of design packages that form the technical specifications and applicable standards for the product. Therefore, this would be the basis for measuring product delivery success, which the team can assess by reviewing the final product and production capacity and comparing them to the specifications. If they align, then the team can claim success in delivering the product.

Project Management Success

The PMP will establish the various performance metrics that the team should achieve. The cost and schedule basis would be taken from the final class 3 estimate. At project close, the organization can determine if the team delivered the project within the set parameters and metrics. If they did, then project management success.

Project Delivery Success

As discussed earlier, this is one of the most difficult dimensions to discuss. Let us consider two scenarios:

- The organization did not specify any measures here; in that case, meeting the first two dimensions is enough to claim success.
- In addition to the first two dimensions, the sponsor added other elements—for example, being able to commission the facility in two months instead of three and without external resources. If this target is achieved, great. Another factor could be no disruptions (shutdown of the plant) for more than one day during the first year of operation.

Objectives Success

This measure of success could be very easy to do but will take years after start-up. One scenario would limit this to achieving the established return on investment. Another scenario could be meeting the break-even period.

44.3 Writing and Publishing a Book

44.3.1 Discovery Phase

This phase would be in line with the standard CAMMP™ model. The idea statement will present the concept of the book, its justification, and strategic alignment to the organization that wants to publish the book. In some cases, if this is an individual project, then strategic alignment might not be a factor.

If management approves the idea, there will be a feasibility study to determine the economics of such a project. This would include market demand and supply, resource availability and capability, subject matter expertise, time and cost considerations, among other factors.

If the project is feasible and management considers it a priority, management will authorize the project.

44.3.2 Requirements Stage

The requirements for such a project will include the expectations of the organizations, the theme of the book, key concepts, sub-concepts, publishing requirements, criteria for success, assumptions, constraints, and risks.

44.3.3 Strategy Stage

The next step after alignment on the project requirements is the development of the PMP, which will include a timeline, cost, resources, guidelines for project control and change management, among other factors.

44.3.4 Definition Stage

With an approved PMP, the project will shift to the Definition Stage and development of a project detailed plan (PDP). The PDP will include a detailed layout for the project (book), deciding on publishing options, finalizing the resource requirements, developing the list of reviewers, updating the project estimate (cost and time), taking a decision on the funding source, finalizing the financial analysis, and seeking final approval.

44.3.5 Implementation Stage

For such a project, it would be best to divide the Implementation Stage into two, three, or four substages. The substages would be sequential but with significant overlaps. The substages could be:

1. **Research.** If this project requires research, it would be necessary to start the research for the various concepts that would be incorporated into the book. The research could continue in parallel to the writing substage.
2. **Writing.** The writing part, which would include writing, reviews, and editing, could be in two drafts (two cycles). The output of this substage is the final manuscript.
3. **Graphics.** If the project requires graphic illustrations and artwork, this work would be going on in parallel to writing.
4. **Publishing.** This could vary based on whether the book is self-published or handled by an external publisher. In either case, this substage will include cover design, layout, proofing, final approval, print setting, and printing.

44.3.6 Operational Readiness Stage

For this project, there are no facilities to operate or new hardware to use. Therefore, this might lead us to believe that there is no project Operational Readiness Stage. Well, there is. Writing and publishing a book for commercial purposes means that the organization will be marketing and selling the book. Therefore, operational readiness could include the following:

1. Updating the chart of accounts to accommodate the costs and revenues from this project.
2. Developing a marketing strategy and plan and implementing them. This could include marketing campaigns, advertising, book signing events, and other activities.
3. Training the business team about the book and its content to convert that information into selling value propositions.
4. Developing a dedicated website for the book.
5. Developing additional materials that could support the book or be used in delivering speeches and workshops related to the book.

44.3.7 Project Close Stage

The Close Stage would be like that in any other project and include a reconciliation between actual and plan, lessons learned, and updating the organizational records and project historical data files for future use.

44.3.8 Project Success

This is the same as for the capital investment project—project success is not a stage but is included here as a separate topic for its importance.

Product Delivery Success

A book is not a road, a plant, or a facility; therefore, determining product delivery success is not as easy for a book project as it is for those physical projects. Possible criteria would be meeting the detailed plan in terms of content and covering all concepts and sub-concepts. Another factor would be management's and reviewers' satisfaction with the product.

Project Management Success

This is simply delivering per the PMP in terms of cost and time, with proper control, and minimal (or no) changes.

Project Delivery Success

This dimension is challenging here; for our purposes, let us limit it to meeting the first two dimensions.

Objective Success

It was stated earlier that the driver for this book is commercial gain. However, other factors would be recognition and spread. To measure this success, possibly a year or two after publication, one could look at:

- Meeting the expected sales revenue and profit margins.
- Getting a few reviewers to offer mostly favorable reviews of the book.
- Garnering readers' feedback satisfaction through surveys, although this could be limited because we are not likely able to know all who read the book.

44.4 Closing Comments

With these samples, we close this chapter.

There could be other samples, and it would be good to show real case studies of completed projects per The Customizable and Adaptable Methodology for Managing Projects™ Model, which would be future projects.

Chapter 45

Closing Chapter

45.1 Introduction

In closing this book, the author cannot and does not claim to know better than the volunteers working with the various professional associations. The book does not critique any of them; they all have excellent products, of high value, that have been in use for many years and decades.

Most of the gaps in these works are by design, as each of these associations has chosen an area to focus on: PMI on process, IPMA on competence, and so on. Therefore, things like a methodology are not part of their mandates. Furthermore, many of the challenges raised are not due to errors by these associations but to misunderstandings by practitioners.

These statements are important to stress and repeat here, because these associations' products, guides, and standards are core to the methodological approach. The CAMMP™ three-dimensional model builds and integrates the learning from these various associations, as well as others that could not be mentioned in detail. However, we did not limit or restrict this book to their platforms, because CAMMP™ offers new ideas and approaches in regard to project management functions, processes, project success, and the tailoring and classification models.

> CAMMP™ offers new ideas and approaches in regard to project management functions, processes, project success, and the tailoring and classification models.

45.2 The Author and Basis for the Book

What is offered here is the perspective of one person who has been working on projects and in project management since 1984, with PMI since 1997, and exposed to IPMA, GAPPS, and AACE, also for many years.

The author has been privileged and had the great opportunity of working on projects small and large, simple and complex. He has worked in different roles (estimating, control, services manager, project manager), contracting scenarios (reimbursable cost plus, fixed price, lump sum turnkey, incentives), and in many countries and most continents. Some of these projects were under an alliance/partnering multi-year contract covering a geographical area with five industrial facilities. Others were mega projects by global joint ventures.

As a consultant, the author has worked with numerous industries covering domains such as government administration, airlines, financial institutions, the military, construction, technology, food, and even jewelry design and manufacturing.

This book is by a practitioner for practitioners!

What we have tried to do here is to reflect back on all of these roles and experiences, from a highly mature project management industry (petroleum), and from leading organizations (Exxon, Shell, Saudi Aramco). These reflections resulted in the research and development effort that has produced The Customizable and Adaptable Methodology for Managing Projects™, The Seven Elements of Project Management Maturity™, The Four Dimensions of Project Success™, and the Four Control Reference Points™.

The results of all the above are based on real-life examples and experiences—not on unproven theories, purely academic research, or literature search.

It is a book *by* a practitioner *for* practitioners.

45.3 Objectives

To remind the reader of the objectives listed earlier in this book: The objectives are to offer individuals and organizations a new way of thinking, a simplified yet effective approach for *applying the principles of project management on real projects in the real world*. The new thinking integrates the following concepts:

- Emphasizing that the process groups are not the project life cycle, and they are not the project phases.
- Offering a new perspective on the process groups, specifically the planning process.
- Emphasizing that standards and guides from professional associations must be part of a holistic organizational project management system—not a standalone resource.
- Offering ideas on how to apply principles of global standards in the real world.
- Introducing a proposed three-dimensional methodological approach for managing projects.

The ultimate objective is to help project management practitioners enhance their own, and their organization's, project performance.

45.4 A New Perspective or Shifting Paradigms?

This book offers a different viewpoint. Some content may even challenge the conventional wisdom for some readers and may lead to shifting their paradigm. Some might challenge what is presented or even dismiss it. The chapters on splitting planning, which changes the established and deep-rooted belief of the five process groups into six, are an example of things that challenge the way we think and work. Presenting more than 60 processes will certainly be a shock to many. The principle of the repeated processes (process groups), which is not new—it has been in the *PMBOK® Guide* since the first formal edition in 1996—but it is the most prominent area of confusion in the project management community today.

The core message and objectives are for practitioners to carefully read what is proposed here, learn about the current practices, acknowledge the various professional associations, and think outside the box. CAMMP™ is based on a professional subject matter expert's opinion gathered from close to three decades of using the concepts of these guides and standards in real-world practice on small, large, and mega projects. These concepts have been applied in industrial projects and on day-to-day internal projects—projects such as writing and publishing a book, launching a new office, launching a business, developing a web portal, building a petrochemical facility, a research and development program, an online application project, and many other projects.

45.5 The Bottom Line

The bottom line:

Growth and continual improvement of any system cannot happen without developers and practitioners being open to addressing weaknesses, gaps, and challenges.

Organizations must discover the power of project management and utilize it to serve them on their journey of excellence. The journey starts with the proper organizational project management system, which includes standardized processes and a systematic, methodological approach.

At this point, it would be appropriate to refer to the CAMMP™ Principles and Critical Success Factors:

- Follow a stage-gate process.
- Incorporate industry leading practices.
- Capture lessons learned throughout project.
- Define roles and responsibilities for all teams.
- Integrate a project life cycle with processes.
- Consider product and delivery processes.
- Customize and adapt to industry and project type.
- Tailor to a given project classification.
- Integrate competence and sustainability.
- Link methods to organizational system.
- Implement project success model with benefits realization.

45.6 The Future

45.6.1 Asset Management

CAMMP™ is a methodology for managing—and—delivering projects. It covers projects end to end, from idea to closure and beyond. However, it is not an asset management model. Ideally, organizations with facilities to operate should follow an asset management process in which the organizational system follows an asset (facility) from concept to market withdrawal (cradle to grave). For these organizations, CAMMP™ would be part of the puzzle, covering the acquisition phase of the product life cycle.

45.6.2 Sample Projects

One can realize that some practitioners might challenge what this book offers. Some might argue that it offers too much, and it is like driving a small nail with a sledgehammer. Others may make a case that unless we cover asset management, CAMMP™ and similar methodologies would fail. In other words, there will always be professionals from both extremes.

The answer to these professionals is here.

This book and the solution we offer make it clear that CAMMP™ is not about a set-in-stone, one-size-fits-all approach. It has to be tailored. Some can use a very basic form of CAMMP™, as explained in Chapter 41, or expand it to the most complex projects. The intent is to publish more books providing case studies, showing various projects types and classes to demonstrate the use of this approach. One book has already been published[1] to present a simple project. Another book, not yet published as of this writing, is on the use of CAMMP™ on a mega (large-complex) project.[2] These future projects can present samples and case studies to help practitioners build their organizational project management systems.

45.6.3 OPMS

As stated in the earlier part of this book, a method, methods, or methodological approach is not enough on its own for excellence in project management. It has to be part of an organizational project management system (OPMS) that is practical, comprehensive, and sustainable. An effective OPMS must be the foundation for organizational excellence in project delivery and the implementation of the strategic plan.

Chapter 9 presented the Seven Elements of Project Management Maturity™ starting with the outer circle—the strategic aspects; as per Figure 9.1. This element can be expanded to include how project management, in the wider context, can be a critical link and a vital area of expertise to implementing the organizational strategy. Furthermore, this element is also about setting the right policies and governance structure for managing portfolios, programs, and projects.

The core of the seven-elements model is the CAMMP™ first and second dimensions, whereas the third dimension is shown in the circles of Figure 9.1 as an advanced topic.

---

[1] *Applied Project Management*, published in 2015 by AuthorHouse.com
[2] Planned for 2018.

In order not to get deeper into this subject, suffice it to say that without an OPMS with all of the elements and organizational records, the benefit of using a methodology cannot be at the level that mature organizations aspire to, if they want to reach the highest level of maturity and center-of-excellence recognition.

45.7 Author's Final Statement

We trust that in this book we were able to demonstrate how to improve the practice of project management and how to apply essential concepts in the real world.

We believe that most would agree that there are numerous valuable points here that must be considered. Ignoring them will not serve the members, certificate holders, and the professional community at large.

We must move beyond the ongoing debates of Scrum versus Kanban, PMI versus IPMA, ISO 21500 versus the *PMBOK® Guide,* and traditional versus adaptive. Project management must move beyond Waterfall and Agile. *This is a challenge to organizations as we urge them not to depend on a single source, regardless of how popular that source is. Popularity is not indicative of excellence!*

We must think out of the box, without ignoring the box!

Appendix A

The CAMMP™ Phases

Project Phase	Phase Main Focus
Discovery	The Discovery Phase is for the team to understand and validate the concept for the project and whether it is feasible.
	It is the initial phase and consists of two stages.
	Note: In guides such as the *PMBOK® Guide* and ISO 21500, this phase is considered "pre-project" and is not part of the project life cycle.
Development	The Development Phase is necessary for developing the concept of the authorized project to ready it for the Delivery Phase. It includes defining the project requirements along with the project management plan and detailed project definition.
	This phase consists of three stages.
Delivery	The Delivery Phase deals with delivering the project in accordance with a detailed plan. Therefore, it includes Implementation and Operational Readiness, along with Initial Operations. It also includes the Close Stage.
	Project delivery encompasses four stages.

Appendix B
The CAMMP™ Stages

Project Stage	Stage Main Focus
Concept	**The Concept Stage is part of the Discovery Phase**
	The concept stage includes documenting the *idea* for the project, outlining the conceptual business case (justification), determining alignment to strategic objectives, and if all are acceptable *and* a priority, management approving the *idea* and authorizing a feasibility study.
Feasibility	**The Feasibility Stage is part of the Discovery Phase**
	The main purpose of this stage is to validate the *idea* and to authorize the *project*. Note the intentional shift from *idea* to *project*, which is to emphasize that only at the conclusion of this phase is there a project authorized to proceed.
	Validating the idea is accomplished via a comprehensive feasibility study that will consider various factors. The feasibility study is the first formal *and* substantial risk management assessment.
Requirements	**The Requirements Stage is part of the Development Phase**
	As the name indicates, this stage identifies and documents the various requirements, including the technical requirements and product characteristics, in addition to the management and project management needs. Furthermore, this stage addresses the need for exploring the expectations of the various stakeholders and converting them to clear and explicit requirements. The stage concludes with a Stakeholder Alignment Gate, which is critical and helps the team freeze the scope to minimize changes and their considerable impact.

(continued on next page)

Project Stage	Stage Main Focus
Strategy	**The Strategy Stage is part of the Development Phase** This stage emphasizes the need to develop the project management (and control) plan along with the delivery strategy, especially for projects with some level of complexity. In this context, "delivery" represents the upcoming Delivery Phase. Furthermore, in this stage the team will develop other strategies, including procurement and best practices. If there is a need for advanced funding, it would be requested after the completion of the class 2 estimate in this stage.
Definition	**The Definition Stage is part of the Development Phase** The Definition Stage is critical and requires significant effort. The main purpose of this stage is to produce the project detailed plan. This plan should include all the details for the scope of work, deliverables, applicable quality standards, safety-health-environmental laws (if applicable), detailed schedule and cost estimates, along with many other details. It is critical to be as clear as possible with enough detail to allow the efficient delivery of the project. Poor planning has severe consequences and is likely to result in the failure of the project to meet some or all of its objectives.
Implementation	**The Implementation Stage is part of the Delivery Phase** The implementation stage is where most of the effort on the project is exercised. It is about performing all of the work packages per the detailed plan. In terms of project management processes, this is a relatively easy stage, but the complexity is due to the significant effort required, depending on the nature of the project.
Operational Readiness	**The Operational Readiness Stage is part of the Delivery Phase** In parallel to the implementation stage, another team should be working on getting ready for post-completion of the physical work. This operational readiness team will be developing operational procedures, maintenance procedures (if a physical facility), training for staff on the new systems, among numerous other activities. Obviously, if a project's product is not something that will be operational, such as a training project, then this stage is likely to be not required and can be eliminated.
Initial Operations	**The Initial Operations Stage is part of the Delivery Phase** Some facilities projects might need a period that one may call *soft launch* or *commissioning*, which are forms of initial operations. In other words, this would be partial operation or operation at reduced capacity. Other projects, such as organizational change projects, may require an initial launch in one business unit instead of the whole organization. This soft launch allows the team to test the product of the project in a controlled environment to learn from it and fix any gaps before full implementation. This soft launch is also called a *pilot*.
Close	**The Close Stage is part of the Delivery Phase** After all work is complete, and before the team can claim completion of the project, successful delivery requires the proper close-out of the project, with all of the necessary activities required at this time. This would include performing the assessment for project success, at least what is possible at this point. Because a few of the dimensions of project success may not be possible at this time, the handover should include a plan for when those measures could be assessed, along with an evaluation of the realization of benefits.

Appendix C

The CAMMP™ Stage Deliverables

Most of these deliverables are typically the output of a corresponding stage and are listed here in sequence.

Project Phase	Deliverable	Description
Discovery	Project Brief	**Output of Concept Stage**
		The project brief is simply a very short document, a few sentences, that describes (1) what the project is, (2) why it is needed (justification), and (3) how it aligns with an organizational strategic plan.
	Feasibility Study	**Output of Feasibility Stage**
		The feasibility study is primarily about *validating the idea*—that is, determining whether the project is viable. This means the project is justified, and the organization can deliver it successfully and realize the expected benefits.
	Project Authorization Document	**Part of the Feasibility Stage, after approval**
		The project authorization document (PAD) authorizes the project. What the PAD is telling the team is that executive management has a clear intent to take the project to completion, although they are only authorizing the next stage.
Development	Project Requirements Document	**Output of the Requirements Stage**
		The project requirements document (PRD) defines the key characteristics, features, and other provisions of the output of the project. It is essentially the high-level scope of the project.

(continued on next page)

Project Phase	Deliverable	Description
Development (continued)	Project Management Plan	**Output of Strategy Stage** The PMP presents the project management strategy/approach for the project. It defines the various steps and works that the team would perform from that point onward, up to project close. It includes the other high-level strategies as well, such as delivery and procurement strategies.
	Project Detailed Plan	**Output of Definition Stage** The project detailed plan (PDP) provides all of the necessary details for the project, such as the scope definition; detailed quality, safety, environmental planning details; detailed schedule and cost estimate; along with various other functions.
	Request for Final Approval	**Output of Definition Stage** This document could be part of the PDP or independent of it. The request for final approval could be just a short letter or form that the project management completes with significant input from the sponsor to submit to executives (decision makers) for the final approval of the project and the authorization of the necessary funds.
Delivery	Deliverables	**Output of Implementation Stage** There is no specific document for this stage. However, this whole stage is about completing the various deliverables and work packages that are required by the project leading to the ultimate delivery of the final project product or output. There could be numerous documents here, each representing the completion of work packages or other items.
	Handover	The handover is a joint deliverable to Implementation and Operational Readiness stages, since it requires input from both of them. Handover, or provisional acceptance, is the client (operation) initial acceptance of the *product of the project* from the project management team. It is usually an indication that the work is essentially complete, and the client is ready to start using the new product, or at least commission it.
	Final Acceptance	**Output of Initial Operations Stage** Once all of the work is complete and the new project is in use, there is often a final acceptance that will take place sometime after handover—at the end of the initial operations stage, if such a stage exists.

Project Phase	Deliverable	Description
Delivery (continued)	Deliverables	**Output of Implementation Stage** There is no specific document for this stage. However, this whole stage is about completing the various deliverables and work packages that are required by the project leading to the ultimate delivery of the final project product or output. There could be numerous documents here, each representing the completion of work packages or other items.
	Handover	The handover is a joint deliverable to Implementation and Operational Readiness stages, since it requires input from both of them. Handover, or provisional acceptance, is the client (operation) initial acceptance of the *product of the project* from the project management team. It is usually an indication that the work is essentially complete, and the client is ready to start using the new product, or at least commission it.
	Final Acceptance	**Output of Initial Operations Stage** Once all of the work is complete and the new project is in use, there is often a final acceptance that will take place sometime after handover—at the end of the initial operations stage, if such a stage exists.
	Close-Out Report	**Output of Close Stage** The major deliverable in the Close Stage is the project close-out report documenting the result of the project work. This report includes reconciliation of the events that took place during the project development and delivery, in addition to lessons learned, organizational records update, and other organizational requirements.
	Others	There are other deliverables in the Delivery Phase. These may include things such as operation and maintenance procedures, training, safety reviews, lessons learned, project success measurements, among other items.

Appendix D

The CAMMP™ Stage Gates

Each of these stage gates is at the end of a corresponding stage and is listed here in sequence.

Stage Gate	Title/Purpose/Description of the Gate
SG1	**Title: Project Brief Approval** Purpose: To approve the project idea, justification, and ensuring strategic alignment The CAMMP™ Model takes the view that every project must be in line with the organizational strategic direction and objectives.
SG2	**Title: Initial Project Approval** Purpose: To approve the feasibility study and to authorize project At this gate, executive management would review the feasibility study results and accept it, if no modifications are required. If the project is not feasible, management terminates the project. If it is feasible, the project's expected value is compared to others within the organization, and if the project is a priority (higher value than other projects), then executive management would authorize the project.
SG3	**Title: Stakeholders' Alignment** Purpose: To ensure alignment of the stakeholders on project requirements SG3's primary objective is to ensure stakeholders' alignment on the project requirements. In other words, did the project manager understand what is required to deliver this project? Requirements include characteristics of the output and outcome.

(continued on next page)

Stage Gate	Title/Purpose/Description of the Gate
SG4	**Title: Project Management Plan Approval** Purpose: To approve the project management plan and gain advance funding, if necessary SG4 is about executive management approval of the project management plan. For this stage gate, the project manager would submit an updated estimate of the project cost and time, which would allow executive management to verify whether the project is still viable and revisit the go/no-go decision. If required, management will also approve funding for the next stage—this is the advance funding process.
SG5	**Title: Final Approval** Purpose: Approval of the project detailed plan and final approval to proceed with implementation This gate is about the final approval of the project. The cost of the work performed before this point is minimal in comparison to the total project cost. Therefore, on most projects, this gate is crucial, and the author labels it "the point of no return," because, once there is approval, it is likely that the project will proceed to completion.
SG6	**Title: Ready for Handover** Purpose: To assess completion of implementation and readiness for handover This gate is concerned with the review of the project implementation and operational readiness work, to verify the completion of the deliverables and to determine if the project management team is ready to hand over the product to the end user's operation personnel.
SG7	**Title: Completion of Implementation Work** Purpose: Final acceptance of implementation work and release of implementation team This gate is at the end of the Implementation Stage and is concerned with confirming that all of the implementation work is complete and verified.
SG8	**Title: Final Acceptance** Purpose: Final acceptance of the project product This gate is concerned with the final acceptance of the result of the project—the output—clearing the project management team to finalize project closure.
SG9	**Title: Project Closed** Purpose: Official closure of the project and financial accounts This final gate is quite different from all of the other gates, because there is nothing after it. We emphasize the need for the gate to stress the point that the project is not closed until all project deliverables are submitted and approved, and the project close-out report is one of those deliverables.

Appendix E

The CAMMP™ Processes

These are the main processes per CAMMP™, which repeat as necessary in every phase or stage.

Process	Purpose and Description
Authorize	There are four process steps under the Authorize process per CAMMP™, which would apply as needed in every stage or phase. These would be the steps to authorize a stage.
Plan Management	There are seven process steps that would be necessary to complete the stage management plan at every stage. The content of the stage management plan from one stage to another is similar, with some variations.
Plan Details	There are 20 process steps, which are necessary to complete the stage detailed plan for every stage. The content of the stage detailed plan from one stage to another would vary based on the scope of the stage. However, in term of topics, the SDP includes the detailed scope of work, WBS, schedule, cost, quality, and all other functions—but for the given stage, not the project.
Implement	CAMMP™ includes nine process steps in this process, and they are mostly related to completing the work, in addition to managing resources and contracts.
Control	Sixteen control process steps would be active from the start of the stage all the way to the end, closing the stage.
Close	Eight closing process steps would be required to close every stage.

Appendix F

The CAMMP™ Process Gates

These process gates are aligned with the processes and repeat at every stage.

Process Gate	Purpose and Description of the Gate
PG1	The first process gate is required at the end of the Authorize process, which leads to issuing the stage authorization document.
PG2	The second process gate is required for approving the stage management plan.
PG3	The third process gate is needed for approving the stage detailed plan.
PG4	The fourth process gate is to verify that all of the stage work is implemented and the team would be ready to close the stage. However, because the completion of the stage work is related to the stage, this process gate could overlap with the corresponding stage gate. Alternatively, this could be the last gate on a stage, to verify completion of the work before submitting the result to the stage gate.

Appendix G

Frequently Asked Questions

It is not common to have a frequently asked questions (FAQ) section in a book, but it will help the reader find the answers to a few important matters about the CAMMP™ Model in an expedited way. There are answers for all of these questions in the book, but a few of them are summarized here.

1. When in the project life should the team select the customized project life cycle for the specific project under consideration? This customization can happen at two different levels:
   a. **Organizational Level.** When an organization decides to implement a project management methodology, whether it is CAMMP™ or not, they should establish standardized project life cycle models that are adapted to their various requirements. If all the organization's projects are of one type, and they are similar in size, then they need one model to follow. With a diversity of project types and sizes/classifications, they will have more than one model to standardize.
   b. **Project Specific.** If there are standardized models, then the project manager could select the one that is most appropriate for the given project. The formal selection takes place during project management planning, although the project manager may have a predetermined idea about the possible model to use during the discovery phase. If there are no organizational models, the project manager develops and formalizes the project life cycle during the discovery phase but finalize it during project management planning.

2. Is this a fixed/standard methodology for all industries?
   One size does not fit all!
   Many project management concepts are common across industries. Nevertheless, some things are specific to an industry. It is proposed that the thought process and sequence of the stages and gates be quite similar for all sectors; the differences are in the application, terminology, and number of stages and gates.

This is the *customizable* part of the CAMMP™ name.

Is this methodology fixed regardless of the size of the project?

No. Even for the same type of project, there should be differences in how to apply the methodology for different project classifications. For example, a small construction project requires a different approach than a major project or a mega project.

This is the *adaptable* part of the CAMMP™ name.

3. What do you mean when you say "project classification"?

   Project classification is a categorization system that organizations can apply to projects to distinguish them based on established criteria; the criteria must be peculiar to an organization or industry and is not readily transferable. For example, "small" could be a project classification, but small in the petroleum industry could mean $1 million to $15 million, whereas in an internal organizational project $1 million could be a huge project.

4. What are common project classifications?

   Once again, it is industry specific. However, one basic classification system would include three main classes or more; the organization must clearly define them. We could also use a simple questionnaire to help decide on a project class. The project classes per the CAMMP™ Model are:
   a. Small-simple projects
   b. Medium-moderate projects
   c. Large-complex projects
   Other possibilities:
   a. Mega projects: typical in the capital project industry.
   b. Strategic projects: this describes projects that are of major strategic benefits to the organization and typically touch a large part of the organization.

   For further reading on the concept of project classifications, we refer the reader to the work of Professor Lynn Crawford at GAPPS[1] and Chapter 41 herein for the CAMMP™ approach.

5. Isn't every project of strategic importance? What does it mean that the objective of stage gate one (SG1) is to confirm that the project aligns with the organizational strategic objectives?

   Organizations must distinguish between strategic projects and projects that align with the organizational strategy. For example, if the organization is in the restaurant business, expanding a restaurant, or opening another one under the same brand and in a similar environment, is likely to be part of the *organizational strategy,* a project that aligns with an expansion strategy. However, opening a new brand or going into a new geographic area, might be a strategic move with significant complexity. Therefore, it can be classified as a *strategic project.*

6. Can the team delete stages?

   It is important to clarify the difference between a stage, as a time span of the project life, and the objective of the stage.

---

[1] http://www.globalpmstandards.org/main/page_complexity.html

Based on the above, organizations can delete stages and have their models contain fewer than nine stages. However, they should not remove the purpose or focus of the stage. In other words, if they remove a stage, they must combine its scope with the previous or subsequent stage. For example, can one eliminate the Requirements Stage? Yes, as long as one includes the requirements with the preceding or succeeding stage. Finally, by doing so, the organization must recognize that they are increasing project risk.

7. Can the team delete stage gates?

   Same answer as in deleting stages (Question 6).

8. What does a disciplined approach mean?

   A disciplined approach is about two essential concepts:
   a. It is about establishing a formal organizational project management system *and following it.*
   b. Within a project, the stage gate approach is about exercising discipline during the project discovery, development, and delivery. In other words, no project should go through the project life cycle without proper reviews and approvals at specific stage gates.

9. Is not a "disciplined approach" a killer of innovation?

   No! A proper methodology that subscribe to the principle of a disciplined approach leads to an *efficient* process, not a *rigid* process. Rigidity is a killer of innovation. However, the time saved due to an efficient process can provide team members with extra valuable time they can utilize to be innovative in developing the project and product. Refer to Chapter 12 for further reading.

10. Is a stage gate process an indicator of risk aversion? Similarly, is project management about being risk averse?

    A stage gate process is essential for minimizing the threats and maximizing the opportunities for delivering a successful project with a good product and realizing the benefits expected. Therefore, a stage gate process is a form of risk management. However, risk aversion is about an attitude or the organizational tolerance to risks. Because no project can realistically be without risks, a risk aversion attitude could lead to letting go of many projects, which in turn leads to stagnation and ultimately may lead a company out of business. Therefore, it is vital that we manage risks through a stage gate methodological approach.

    Refer to Chapter 12 for further reading.

11. What is project management maturity?

    The term *project management maturity* is linked to organizational project management. The concept is that each organization has a different level of maturity of project management practices. Some may have excellent systems in place, and others may be doing projects in an ad hoc, unstructured, or informal way.

    For example, an organization without any formal documented system on how to manage projects occupies a lower level of project management maturity. An organization with a well-established system, one which team members understand and follow, and that embraces continual improvements can be positioned at a higher level of organizational project management maturity.

There are various project management maturity models on the market; most of them use a scale of one to five to measure the level of maturity within organizations. PMI has a system that is not numerical, but is based on the concept of SMCI—Standardize, Measure, Control, and Improve. SUKAD has also developed a basic model that consists of seven elements of project management maturity[2]; however, the SUKAD model is not designed for assessment but to help organizations build a sustainable organizational project management system.

12. Is the stage gate the same as a milestone?

No, although they may coincide.

Some use the term *milestone* for achieving a certain objective, which could be the same thing as a stage gate approval. However, the definition of a milestone is a flag of an important date on the project schedule or life cycle. Furthermore, in scheduling, one defines a milestone as an activity with zero duration and zero resources.

For example, the *date* of approving the project at SG5 is a milestone. The stage gate itself is more than a fixed date; it is a process requiring input (the work from the previous stage), reviews, and debates leading to a decision. None of this can be accomplished in zero time and with no resources.

Once again, the stage gate is a process leading to a decision. The date of the decision is what we "flag" as a milestone.

13. What is operational readiness?

Any project that requires the handover of a product or something similar to an internal or external client and that requires the client to manage it or operate it will likely require preparation before the handover of the product to the customer.

The project could be a new employee performance management system that the human resources department is implementing. Before the new system goes live, the employees of the HR department, in addition to supervisors and staff across the organization, need to learn about the system.

Similarly, in an industrial plant, operation and maintenance policies and procedures have to be in place before initial operations.

14. Is operational readiness required for all projects?

No. Some projects would not require operational readiness. For example, demolishing a building, there is nothing to operate once completed.

15. How do you link CAMMP™ to the *PMBOK® Guide* or ISO 21500?

The *PMBOK® Guide* and ISO 21500 both offer a set of processes that are linked to 10 knowledge (subject) areas and five process groups. These processes repeat in every

---

[2] http://sukadway.sukad.com/

stage per these leading guides, but many practitioners miss this point and assume that the *process groups* are *project stages*. CAMMP™ is a three-dimensional model that is founded on the first dimension—a project life cycle approach. A project life cycle consists of phases and stages, and to manage each phase or stage, a project team must utilize a set of processes, which is the CAMMP™ second dimension. In other words, the processes of the *PMBOK® Guide* and ISO 21500 are a subset of CAMMP™.

Please refer to Part D for further reading.

16. **Is the intention of this methodology to be a replacement for the guides and standards published by professional associations?**

    No! The various professional associations offer useful guides and standards, each with a different focus. What CAMMP™ does is integrate the knowledge of these diverse resources into a universal, three-dimensional, practical model.

    CAMMP™'s aim is to complement and supplement these guides.

Acronyms

SUKAD/CAMMP™–Specific Terms	
CAMMP™	Customizable and Adaptable Methodology for Managing Projects™
COR	Close-out report
CPM	Copy-paste-modify
CRP	Control reference point
FS	Feasibility study
OPMS	Organizational project management system
PAD	Project authorization document
PDP	Project detailed plan
PG	Process gate
PMP	Project management plan
PB	Project brief
PRD	Project requirements document
SAD	Stage authorization document
SDP	Stage detailed plan
SG	Stage gate
SMP	Stage management plan
SUKAD	Success Uniqueness Knowledge Attitude Development (the name of the company co-founded by the author)
All Other Terms	
AACE®	The Association for the Advancement of Cost Engineers, International (AACE® International)
ANSI	American National Standards Institute
BP	Best practices

All Other Terms (continued)	
CII	Construction Industry Institute
CPM	Critical path method
EPT	Extended project team
GAPPS	Global Alliance for Projects Performance Standard
GPM®	Green Project Management
ICB®	Individual Competence Baseline (IPMA)
IPMA	International Project Management Association
ISO	International Organization for Standardization
NGO	Non-governmental organization
NPO	Nonprofit organization
OCB®	Organizational Competence Baseline (IPMA)
OPM	Organizational project management
PBS	Product breakdown structure
PEM	Project Excellence Model (IPMA)
PLC	Product life cycle/project life cycle
PM	Project manager
PM	Project management
PMBOK® Guide	*A Guide to the Project Management Body of Knowledge®*
PMI	Project Management Institute
PMP	Project management plan
PMP®	PMI® Project Management Professional certification
PMT	Project management team
PRiSM™	Projects integrating Sustainable Methods™ (GPM®)
PT	Project team
RFP	Request for proposal
SOW	Scope of work / statement of work
VIP	Value-improving practices
WBS	Work breakdown structure

Glossary of Terms

Term	Acronym	Description
Best practices	BP	*See* value-improving practices
Challenged project		A challenged project is one that is completed but possibly missing one or more objectives and possibly experienced issues in some of the project performance metrics, such as cost or schedule.
Control reference points	CRP	These are points across the project life cycle that are used for project control and measuring project success. CRPs are also used within a stage. *(Refer to Chapter 35.)*
Deliverable		A defined work item (scope) that, when finished, is delivered for review/inspection and approval.
Extended project team	EPT	The extended project team includes the project team (project manager, PMT, and functional/technical professionals) and the operational readiness team.
Failed project		A failed project is one that the organization terminates before completion, or is completed with major issues; in other words, it did not achieve the original objectives.
Gate		*See* stage gate *or* process gate
Management		In various places in the book, the term *management*, unless noted otherwise, refers to executive management, which could be a single person, such as owner, CEO, general manager, project sponsor. Alternatively, this term can be a reference to an executive committee, a steering committee, a board of directors, or any other formal management group.
Method		"A particular procedure for accomplishing or approaching something, especially a systematic or established one."[a]

[a] Oxford Dictionaries (http://oxforddictionaries.com/)

Term	Acronym	Description
Methodology		"A system of methods used in a particular area of study or activity."[b]
One size fits all		In the context of this book, this term is not limited to size. It is used to represent the differences between projects in term of size, complexity, domain, function, or types of organizations involved in the projects. In other words, projects might share project management processes, but one cannot use the same method for all kinds of projects.
Phase		This term is often used interchangeably with *stage*, but in this book and the CAMMP™ Model, we differentiate between these terms and use *phase* to indicate the three principal phases of a project, which are Discovery, Development, and Delivery.
Pilot		A pilot, in the context of this methodology and book, means *experimental period*, wherein the project could start with a trial period. For example, an organization is embarking on changing their performance management system. After the completed design of the system, they might choose to roll it out in one department before all others. The pilot is done to capture the learning in a controlled environment. *(Refer to Chapter 25.)*
Process gate	PG	Equivalent to an interim review, control point, or decision point but within the stage. Each process gate occurs at the end of some of the given processes. In the standard CAMMP™ Model, there are four process gates per set of processes. *(Refer to Chapter 16.)*
Product		In the context of this book, unless otherwise noted, the word *product*, especially the project's product, refers to the final output of the project delivery, such as an industrial facility, hospital, book, software application, or new organizational system.
Product breakdown structure	PBS	Product breakdown structure is similar to the work breakdown structure, except the WBS covers scope (project work) items only, and the PBS incorporates a breakdown of the product's components.
Project authorization document	PAD	The PAD is a document to announce management official permission to authorize the project. *(Refer to Chapter 19.)*
Project detailed plan	PDP	The Project Detailed Plan is the output of the Project Definition Stage and includes details about the project various components, such as scope, quality, time, risk. *(Refer to Chapter 22.)*

[b] Ibid.

Glossary of Terms

Term	Acronym	Description
Project life cycle	PLC	PLC is the sequential timeline of the project phases and stages from idea to closure.
Project management team	PMT	The project management team is the team supporting the project manager in managing the project and includes cost, schedule, quality personnel, and others as necessary (safety, construction, logistics). The project management team includes the project manager.
Project owner		This term refers to the organization that is developing and will own the product of the project.
Project requirements document	PRD	The project requirements document is the deliverable that outlines the key provisions of the project, as the project manager and team understand them. *(Refer to Chapter 20.)*
Project team	PT	The project team includes the technical (or functional) professionals who will perform the detailed work of the project (designers, engineers, specialists). The project team in the wider context includes the project manager and other project management team members.
Project's product		The *PMBOK® Guide* defines the output of the project as a product, service, or result. A product is a physical facility (building, refinery); a service would be a project that delivers a service (training, consultancy); a result is the output of research and development that others could use to provide a service or further research and development. In this book, we use this term often generically to refer to the product, service, or result.
Responsibility assignment matrix	RAM	Defines and clarifies the roles and responsibilities of the people involved in the project.
Stage		A defined time span and part of the project life cycle that usually ends with the production of significant deliverables. In the generic CAMMP™ Model, there are nine stages: concept, feasibility, requirements, strategy, definition, implementation, operational readiness, initial operations, and close. *(Refer to Chapter 13.)*
Stage deliverable		Same as a deliverable. However, stage deliverables are the main deliverables, usually approved at a stage gate.
Stage gate	SG	Equivalent to an interim review, control point, or decision point. Each stage gate occurs at the end of a project stage. In the standard CAMMP™ Model, there are nine gates.
SUKAD Model		The name used alternatively to refer to the SUKAD methodology for managing projects, CAMMP™.

Term	Acronym	Description
SUKAD Way™, The		The name used to refer to the SUKAD approach for managing projects, to encompass the SUKAD Model and the other products from its research and development program, including *The Seven Elements of Project Management Maturity™*.
Value-improving practices	VIP	Value-improving practices deal with practices that are not commonly used, and if organizations use them properly, they can expect significant performance improvement in the areas of safety, quality, cost, schedule, performance, among other areas. VIP should be limited in number because they are often differentiating factors between typical performance and best-in-class performance. *(Refer to Chapter 31.)*
Work breakdown structure	WBS	Work breakdown structure is the logical breakdown of the project scope into smaller and smaller components, reaching the work packages level. The breakdown is necessary to facilitate planning, to estimate cost and time, and for control of the project.
Work package		A work package is a WBS element that is at the bottom of a given branch—the lowest level. Since it is a WBS element, then it is a deliverable, a specific scope of work.

Bibliography

AACE® International (n.d.). *Cost Estimate Classification System.* Sample PDF available at https://web.aacei.org/docs/default-source/toc/toc_18r-97.pdf?sfvrsn=2

Ajam, M. A. (2006). Awakening the Giant Within. PMI Europe-Middle East-Africa Congress (EMEA) 2006. Madrid: The Project Management Institute.

Ajam, M. A. (2010). *The Inheritance.* Dubai, United Arab Emirates: SUKAD.

Ajam, M. A. (2014). *Project Management Foundation.* Amioun, Al-Koura, Lebanon: SUKAD Multimedia.

Ajam, M. A. (2014). *Redefining the Basics of Project Management.* Amioun, Al-Koura, Lebanon: SUKAD Multimedia.

Ajam, M. A. (2015). *Applied Project Management.* AuthorHouse.

Construction Industry Institute (2016, September). Knowledge Base from CII. Retrieved from https://kb.construction-institute.org/Best-Practices

Englund, R. L. (2006). *Project Sponsorship.* San Francisco, CA: Jossey-Bass.

Gallup (2016, October 30). Retrieved from http://www.gallup.com/businessjournal/152429/cost-bad-project-management.aspx

GPM Global (2013). *The GPM® Reference Guide for Sustaiability in Project Management.* Information available at https://greenprojectmanagement.org/the-gpm-reference-guide-to-sustainability-in-project-management

GPM Global (2016). *The GPM® Global P5™ Standard for Sustainability in Project Management.* GPM Global. Information available at https://greenprojectmanagement.org/the-p5-standard

IPMA (2015a). ICB®: IPMA Competence Baseline. Retrieved from International Project Management Association: http://www.ipma.world/certification/competence/ipma-competence-baseline/

IPMA (2015b). *Individual Competence Baseline for Project, Programme, and Portfolio.* International Project Management Association. Available at http://products.ipma.world/wp-content/uploads/2016/03/IPMA_ICB_4_0_WEB.pdf

ISO (n.d.). ISO 21500 Guidance on Project Management. ISO. Information available at https://www.iso.org/standard/50003.html

Mounir A. Ajam and SUKAD Group (2012–2014). SUKAD Group Blog Site. Retrieved from Redefining Project Management: http://blog.sukad.com

Peter Principle. Retrieved April 26, 2014, from http://www.investopedia.com/terms/p/peter-principle.asp

Project Management Institute. (2013a). *A Guide to the Project Management Body of Knowledge (PMBOK® Guide)*, 5th ed. Newtown, IL: Project Management Institute, Inc.

Project Management Institute (2013b). Thought Leadership Series. Available at https://www.pmi.org/learning/thought-leadership/series/pmo

Project Management Institute. (2017). *A Guide to the Project Management Body of Knowledge (PMBOK® Guide)*, 6th ed. Newtown, IL: Project Management Institute, Inc.

SUKAD Group (2016). Project Management. iSMILE™ Series, Video 5, *PM, Idea Mgmt Strategic Alignment.* Accessible at https://youtu.be/MPZxI815nP8

W. Edwards Deming Institute. Retrieved January 21, 2017, from https://deming.org/explore/p-d-s-a

Axelos (n.d.). What Is PRINCE2? Axelos Global Best Practice. Retrieved August 21, 2015, from https://www.axelos.com/best-practice-solutions/prince2/what-is-prince2

Index

A

AACE. *See* Association for the Advancement of Cost Engineers International
acceptance, 34, 45, 59, 100, 128, 135, 154, 161, 163–165, 167, 169, 171, 172, 187, 189, 190, 218, 270, 288
acceptance criteria, 100, 128
acceptance, final, 154, 161, 163, 165, 167, 169, 171, 187, 218, 288
accept custody, 161, 165, 218
adaptive approach, 267
adaptive principles, 62
advanced topics, 53, 134, 179, 182
advance funding, 141, 217, 222, 223, 227, 286
Agile, 5, 8, 15, 30, 46, 61, 68, 77, 152, 239, 253, 258, 297
Agile principles, 61, 68
Agile/Scrum, 46, 61, 239
Agile versus Waterfall, 258
American National Standards Institute (ANSI) 11, 15, 19
ANSI. *See* American National Standards Institute
applied project management, 90, 296
applying CAMMP™, 257
applying the principles of project management, 26, 294
approval, final, 77, 140, 142, 144, 151, 154, 159, 218, 219, 222–224, 238, 283, 289, 290
approvals, timing of, 219
approvals, types of, 217, 223
a project example, 213
asset management, 176, 184, 296
Association for the Advancement of Cost Engineers International (AACE), 8, 52, 57, 195, 225, 226, 293
Authorize Process, 95, 100, 102, 190, 207, 218, 233, 234, 265

B

baseline, 4, 8, 21, 46, 56, 60, 66, 140, 155–157, 159, 193, 195, 227, 233, 235–237
benefits realization, 43, 45, 60, 89, 171, 173, 175, 177, 187, 295
benefits realization evaluation, 173, 175
best practices, 13–15, 19, 53, 81, 87, 128, 130, 134, 182, 201, 202, 219, 220, 240, 258
best practices and CAMMP™, 202

327

best practices strategy, 134
best practices, value of, 202
budget approvals, 218, 221–224, 228
budget, constrained, 219, 220
budgeting common practices, 220
business case, 34, 67, 69, 76, 107, 114, 115, 207, 235
bus trip analogy, 249–251

C

CAMMP™ principles, 295
CAMMP™ processes, 94, 99, 102, 233
CAMMP™ three-dimensional model, 177, 293
capital investment project, 285, 290
challenges in current practices, 23
change management, 46, 69, 102, 128, 136, 165, 201, 202, 238–240, 242, 243, 265, 289
change management in the fog, 240, 242
changes, types of, 46, 243
characteristics, 34, 58, 77, 125, 126, 146, 220
CII. *See* Construction Industry Institute
class 1 estimate, 140, 214, 226–229, 231, 237, 286
class 2 estimate, 140, 142, 148, 227, 231, 237, 241, 242, 286
class 3 estimate, 140, 227, 229, 231, 241, 242, 288
classification in project management guides, 263
classification system, 40, 225, 261–263
Close Process, 102, 103, 172, 190, 214
Close Stage, 37, 38, 61, 69, 79, 80, 99, 103, 171–173, 177, 186, 212, 216, 231, 244, 248, 274, 288, 290
closing, 6, 9, 12, 15, 17, 19, 22, 30, 35, 38, 40, 48, 54, 60–62, 65, 69, 72, 81, 88–90, 92, 93, 95, 97, 103, 111, 146, 161, 168, 171, 172, 175, 177, 186, 190, 208, 216, 218, 223, 231, 244, 246, 251, 259, 267, 270, 275, 283, 285, 291, 293
competence, 4, 8, 21, 22, 25, 44, 52, 53, 56, 57, 65, 66, 86, 87, 90, 162, 182, 193–196, 247, 248, 293, 295
competence and CAMMP™, 196
competence in organizations, 195
Concept Stage, 76, 107, 109
Construction Industry Institute (CII), 8, 87, 128, 130, 134, 201, 240
controlling, 17, 18, 30, 35, 37, 38, 48, 72, 94, 107, 130, 154, 237, 246, 267
Control Process, 99, 102, 103, 107, 214, 234, 265
control reference point (CRP), 38, 160, 234–237, 243
critical challenge, 29
critical success factors, 295
CRP. *See* control reference point
customization, 43, 44, 64, 277, 279–282
customizing and adapting CAMMP™, 277

D

Definition Stage, 77, 89, 124, 135, 137, 143, 151, 155, 160, 202, 215, 222, 223, 227, 237, 241, 242, 265, 273, 286, 288, 289
deliverables within a stage, 99
Delivery Phase, 60, 69, 74, 77, 78, 80, 123, 131, 135–138, 147, 153, 163, 167, 171, 173, 202, 208, 211, 215, 237, 250, 267, 275
delivery strategy, 77, 78, 131, 133, 134
description of the product, 110, 121, 125
detailed planning, 46, 69, 90–92, 102, 132, 135, 142, 146, 151, 152, 210, 212, 219, 241, 246, 273
Development Phase, 74, 77, 121, 123, 128, 129, 131, 143, 202, 208, 210, 214, 236, 238, 244, 266, 286
disciplined approach, 59, 67, 70, 228, 267, 278, 282

Discovery Phase, 74, 76, 80, 89, 107, 108, 113, 121, 164, 176, 208, 209, 213, 226, 238, 241, 244, 246, 250, 266, 272, 273, 275, 286, 289

E

environment, 5, 8, 14, 18, 22, 55, 61, 64, 79, 88, 95, 97, 117–119, 152, 157, 158, 162, 174, 182, 194–196, 198, 253, 263, 281
EPT. *See* extended project team
estimate, class 1, 140, 214, 226–229, 231, 237, 286
estimate, class 2, 140, 142, 148, 227, 231, 237, 241, 242, 286
estimate, class 3, 140, 227, 229, 231, 241, 242, 288
estimate classification, 228, 230
estimate, components of good, 228
estimating, 13, 14, 63, 69, 140, 148, 149, 216, 223, 227–231, 241, 265, 294
estimating and project management maturity, 229
estimating and the project life cycle, 230
expectations, 77, 102, 124–126, 129, 130, 140, 144, 158, 162, 227, 273, 280, 283, 289
expected outcome, 110, 119
extended project team (EPT), 138, 139

F

Feasibility Stage, 76, 107, 113, 140, 145, 160, 207, 208, 213, 214, 231, 247, 248
feasibility study, 39, 47, 65, 69, 76, 78, 107–109, 113–116, 118–122, 124, 128, 140, 141, 145, 150, 162, 175, 188, 190, 198, 207, 208, 210, 213, 214, 216, 217, 226, 230, 236, 237, 246, 247, 250, 270, 272, 280, 286, 289
feasibility study and risk management, 115
financial considerations, 119

Four Control Reference Points™, 237, 239, 240, 294
framework, 4–7, 14, 51, 52, 193, 195, 196, 281, 282
front-end planning, 128, 202

G

GAPPS®. *See* Global Alliance for Projects Performance Standards
gaps and opportunities, 25
gaps in practice, 29, 181, 183
Global Alliance for Projects Performance Standards (GAPPS®), 8, 56, 87, 193, 196, 261, 293
global associations, 7, 8, 55, 86
good practices, 13, 14, 201, 217, 283
governance, 6, 54, 69, 73, 75, 77, 80, 141, 176, 183, 191, 208, 220, 229, 240, 278, 279, 296
GPM®. *See* Green Project Management Global
Green Project Management Global (GPM®), 5, 8, 22, 25, 45, 51, 55, 57, 62, 66, 87, 118, 128, 134, 197, 198
Guide to Sustainability in Project Management, 198

H

handover, 79, 154, 161, 163–167, 169, 218, 288
health, 97, 118, 135, 138, 144, 146, 147, 152, 158, 174, 189, 190, 213, 253
health, safety, and environment (HSE), 97, 101, 102, 138, 146, 147, 152, 151, 174, 194
hidden changes, 240, 242, 244
high-level scope, 145
history of CAMMP™, 73
how to close the project, 134, 137
how to control the project, 134, 136
how to customize, 280

how to define the project, 134, 135, 146, 158
how to develop proper estimates, 148
how to get ready for operation, 136
how to manage change on the project, 134, 136
how to manage the project, 132, 134, 135, 155
"how-to" section, 134
HSE. *See* health, safety, and environment

I

ICB®. *See* Individual Competence Baseline
idea statement, 38, 76, 107, 189, 234, 272, 289
Implementation Stage, 78, 113, 130, 135–137, 144, 150, 153, 154, 159–161, 163, 165, 171, 174, 212, 215, 227, 272, 274, 287, 289
implementing, 17, 18, 30, 35, 37, 38, 69, 88, 94, 103, 134, 160, 164, 168, 193, 195, 201, 208, 210, 216, 290, 296
Implement Process, 101, 102, 160, 210, 214, 234
importance of proper project closure, 172
Individual Competence Baseline (ICB®), 8, 21, 25, 56, 66, 193, 194, 263
initial operations, 37, 47, 60, 78, 79, 137, 153, 162, 163, 165–167, 169, 190, 202, 212, 215, 218, 271, 279, 288
Initial Operations Stage, 78, 163, 167, 215, 218, 279, 288
initiating, 17, 30, 32, 35, 36, 88, 89, 108
innovation, 57, 70, 71, 109, 281
International Organization for Standardization (ISO), 4, 6, 9, 12, 17–19, 22, 25, 26, 29–32, 38, 40, 43, 44, 46, 48, 52, 55, 56, 58, 59, 66, 76, 80, 85–88, 90–93, 95, 97, 120, 131, 134, 138, 140, 141, 148, 160, 171, 181, 198, 201, 205–207, 233, 234, 245, 248, 263, 297

International Project Management Association (IPMA), 4, 7–9, 17, 21, 22, 25, 43, 44, 51, 52, 55, 56, 59, 62, 66, 86, 87, 134, 193–197, 293, 297
IPMA. *See* International Project Management Association
ISO. *See* International Organization for Standardization
ISO 21500, 4, 6, 12, 17–19, 22, 25, 29–32, 40, 43, 46, 48, 52, 56, 66, 76, 85, 87, 88, 90–93, 148, 198, 201, 205, 207, 233, 263, 297

J

justification(s), 48, 76, 107, 110, 114, 115, 117, 120, 121, 133, 188, 236, 242, 263, 267, 289

K

key features of CAMMP™, 58
knowledge areas, 6, 7, 11–13, 17, 18, 25, 31, 46, 56, 87, 88, 93, 264

L

large-complex projects, 74, 212, 261, 266
leading practices, 51, 72, 85, 201, 220, 225, 295
lessons learned, 53, 102, 103, 120, 171–177, 216, 229, 230, 251, 274, 283, 290, 295
life cycles 360°, 269
link to global standards, 85

M

management and detailed planning, 91, 210
management planning, 46, 90, 92, 140, 150, 151, 208, 217–219
managing across the stages, 205, 207, 208
managing per the project's class, 261
managing risks, 245, 248

managing risks, timing for, 248
market evaluation, 117
medium-moderate projects, 261, 265
mega project, 71, 127, 134, 137, 147, 149, 202
method, 4–6, 9, 14, 15, 19, 22, 25, 30, 44, 51, 52, 54, 56, 57, 80, 87, 90, 97, 133, 134, 152, 198, 202, 277, 278, 296
methodology, 4, 5, 9, 11, 14, 15, 17, 21, 25, 29, 30, 43, 44, 48, 54, 56–61, 64, 65, 70, 72, 73, 75, 85, 90, 99, 103, 135, 166, 171, 176, 181, 190, 193, 229, 239, 253, 258, 259, 269, 275, 278, 280–283, 291, 293, 294, 296, 297
mobilize project manager, 100, 102, 213
monitor and control, 48, 237, 246
myths about the *PMBOK® Guide*, 13

N

new perspective, 26, 27, 294, 295

O

objective success, 47, 100, 189, 190, 291
OCB®. *See* Organizational Competence Baseline
offered solution, 9, 17, 26, 27, 49
operational readiness, 47, 60, 78, 135–137, 153, 154, 161, 163, 164, 166, 167, 202, 212, 215, 250–252, 274, 279, 287, 290
operational readiness personnel, 251
Operational Readiness Stage, 78, 135, 163, 164, 167, 212, 215, 274, 279, 287, 290
OPMS. *See* organizational project management system
Organizational Competence Baseline (OCB®), 8, 21, 56, 66, 193
organizational project management, 4–6, 9, 13, 26, 44, 45, 51, 54, 55, 69, 86, 90, 100, 133, 172, 176, 181, 185, 191, 195, 213, 229–231, 237, 244, 253, 258, 264, 280, 283, 294–296

organizational project management system (OPMS), 6, 44, 51, 52, 54, 55, 100, 133, 176, 181, 195, 213, 229, 258, 264, 267, 280, 283, 296, 297
organizational records, 171–174, 176, 177, 216, 274, 290, 297
organizational system, 43, 44, 52, 132, 135, 143, 162, 163, 195, 230, 270, 279, 281, 282, 295, 296
output of the project, 110, 125, 154, 175, 184, 273

P

P5. *See* Standard for Sustainability in Project Management
PAD. *See* project authorization document
PBS. *See* product breakdown structure
PDCA. *See* Plan, Do, Check, Act
PDP. *See* project detailed plan
people aspects, 137, 249, 265
people competences, 22, 194
performance management, 136, 140, 155–157, 227
performance management baseline (PMB), 140, 155–157, 227
personal story, 251
perspective competences, 22, 194
Phases
  Delivery, 60, 69, 74, 77, 78, 80, 123, 131, 135–138, 147, 153, 163, 167, 171, 173, 202, 208, 211, 215, 237, 250, 267, 275
  Development, 74, 77, 121, 123, 128, 129, 131, 143, 202, 208, 210, 214, 236, 238, 244, 266, 286
  Discovery, 74, 76, 80, 89, 107, 108, 113, 121, 164, 176, 208, 209, 213, 226, 238, 241, 244, 246, 250, 266, 272, 273, 275, 286, 289
pilot, 79, 137, 163, 167, 168, 212, 215, 280
Plan Details Process, 101, 102, 214, 234, 265

Plan, Do, Check, Act (PDCA), 18, 48, 90, 93, 181
Plan Management Process, 101, 102, 213, 234, 265
planning, 17, 18, 26, 30, 35–37, 46, 54, 57, 63, 69, 70, 75, 78, 88–92, 94, 101, 102, 108, 109, 114, 128, 129, 132, 135, 137, 139–142, 144, 146, 148, 150–152, 162, 174, 187, 202, 207, 208, 210, 212, 213, 217–219, 222, 227, 229, 230, 238, 239, 241, 242, 246, 247, 273, 287, 294, 295
planning for modularization, 202
planning for startup, 202, 287
PMBOK® Guide, 4, 6, 7, 11–15, 17–19, 29–32, 39, 40, 43–47, 52, 56, 60, 63, 66, 76, 85–93, 97, 126, 143, 148, 194, 201, 205, 207, 233, 246, 259, 263, 264, 281, 282, 295, 297
PMB. *See* performance management baseline
PMI. *See* Project Management Institute
PMP. *See* project management plan
PMT. *See* project management team
portfolio management, 13, 21, 54, 80, 120, 141, 253
potential pitfalls, 109, 117, 119, 222, 281–283
practical management of the three dimensions, 203
practical real-world application, 255
practice competences, 22, 194
practice gaps, 26
PRD. *See* project requirements document
pre-handover, 136, 164, 270
pre-project, 45, 47, 60, 69, 76, 80, 207, 208, 250
preset budget, 220
PRINCE2®, 5, 6, 9, 25, 56, 198
PRiSM™, 5, 6, 8, 22, 45, 52, 66, 87, 134, 198, 263
Processes
  Authorize, 95, 100, 102, 190, 207, 218, 233, 234, 265
  Close, 102, 103, 172, 190, 214
  Control, 99, 102, 103, 107, 214, 234, 265
  Implement, 101, 102, 160, 210, 214, 234
  Plan Details, 101, 102, 214, 234, 265
  Plan Management, 101, 102, 213, 234, 265
  Stage Gate, 68, 70, 218
processes and functions, 7, 40, 52, 55, 56, 85, 87, 93, 177, 181
process gate, 95, 99, 208, 223, 234
process gates approvals, 218, 223
process groups, 6, 11, 12, 15, 17–19, 22, 25, 26, 29–32, 34–36, 38–41, 46–48, 52, 58, 68, 76, 80, 85–90, 93, 97, 140, 172, 193, 196, 198, 205, 206, 215, 217, 246, 258, 267, 275, 294, 295
process groups and project life cycle, 29, 246
procurement, 18, 31, 48, 70, 102, 113, 127, 132, 134, 150, 151, 174, 208, 212, 214, 216, 220, 229, 237, 251, 253, 265, 287
product breakdown structure (PBS), 6, 102, 126, 127, 140, 146, 152
product delivery success, 184, 185, 189, 288, 290
product life cycle, 6, 33, 184, 188, 191, 269–271, 296
product processes, 19, 46, 57
product scope, 6, 92, 132, 145, 146
professional associations, 8, 9, 17, 21, 25, 26, 44, 48, 51, 65, 86, 87, 90, 182, 193, 258, 259, 293–295
professional development, 53, 193, 195
professional societies, 7
program life cycle, 6, 33
program management, 86, 253, 283
project approvals, 69, 109, 152, 154, 217, 227, 230
project authorization document (PAD), 88, 89, 100, 113–115, 120, 121, 123, 124, 128, 158, 160, 163, 186, 206, 208, 214, 221, 236, 241, 243, 248, 272, 273, 275
project boundaries, 126

project brief, 76, 89, 107, 109–111, 114, 121, 124, 129, 136, 143–145, 155, 157, 189, 190, 213, 234–236, 241, 243, 272
project change management, 46, 128, 239, 240, 242, 243
project charter, 31, 88, 89, 114, 120, 206–208, 216
project classification, 5, 44, 55–57, 133, 196, 261, 263, 295
project control, 52, 60, 136, 147, 148, 155, 172, 193, 195, 227, 233–236, 238, 239, 244, 265, 289
project delivery success, 47, 100, 175, 186–189, 288, 291
project detailed plan (PDP), 77, 78, 88, 89, 92, 124, 125, 130, 132, 135, 143–146, 148, 151, 152, 154, 155, 157–159, 160, 165, 169, 173, 184, 186, 206, 218, 237, 238, 241–243, 273, 280, 283, 289
project domain, 34, 56, 62, 81, 114, 139, 147, 190, 249, 278, 279
project estimates, 202, 216, 225, 226, 231
project failure, 46, 162, 202
project life cycle, 5, 6, 14, 15, 22, 25, 26, 29–38, 40, 41, 43, 45–47, 52, 55, 57–65, 67, 68, 74–76, 78–80, 85, 87, 89, 90, 92, 93, 95, 103, 105, 115, 120, 129, 133, 140, 168, 171, 173, 177, 181, 182, 184, 190, 191, 193, 196, 198, 199, 201, 202, 205–208, 218–220, 222, 225, 226, 228, 230, 233, 240–242, 245, 246, 249, 253, 264, 265, 269–272, 274, 275, 277, 278, 281, 285, 286, 294, 295
project life cycle selection, 133
project management functions, 18, 93, 95, 136, 144, 148, 151, 155, 196, 253, 293
Project Management Institute (PMI), 4, 6–8, 11–15, 17, 19, 22, 25, 26, 39, 43–45, 48, 51, 52, 55, 56, 58–60, 62, 80, 86, 87, 91, 93, 95, 120, 131, 134, 138, 140, 141, 143, 171, 181, 206, 234, 245, 259, 284, 293, 297
project management maturity and control, 238

project management plan (PMP), 31, 36, 38, 60, 88, 89, 91, 92, 115, 130–134, 136–141, 143, 146, 147, 150, 158, 186, 187, 189, 194, 199, 206, 210, 212, 215–217, 223, 227, 231, 233, 235, 238, 241, 248, 273, 280, 282, 283, 286, 288, 289, 291
project management process, 7, 19, 22, 47, 52, 57, 58, 78, 83, 87, 90, 93, 94, 109, 126, 177, 190, 206, 208, 213, 229
project management strategy, 36, 77, 133, 134
project management success, 47, 79, 100, 175, 185, 189, 288, 291
project management team (PMT), 47, 48, 77, 101, 113, 114, 118, 119, 123, 129, 131, 135, 137, 138, 140, 144, 147, 148, 150, 161, 165, 166, 169, 174, 176, 184–186, 206, 208, 212–216, 218, 229, 230, 234, 250, 251, 262, 264, 287, 288
project manager, 8, 44, 48, 64, 95, 100–103, 108, 111, 113, 114, 120, 121, 123, 124, 133, 135–139, 144, 153, 155, 159, 161, 162, 164, 166, 172, 174, 183, 184, 187, 189–191, 193, 194, 206, 213, 214, 216, 218, 229, 236, 237, 243, 250, 251, 264, 280–283, 294
project manager authority, 281
project owner, 34, 39, 45–47, 61, 65, 80, 154, 166, 183, 184, 187, 207, 222, 225, 271–275, 282, 287
project planning, 36, 132, 227
project procurement, 134, 150
project procurement strategy, 134
project requirements document (PRD), 77, 123–125, 127, 131, 143, 146, 151, 214, 215, 223, 237, 241, 273, 286
project risk management, 245
project samples, 64, 258
project scope, 6, 92, 126, 127, 132, 143, 145, 146, 149, 220, 230, 238
project sponsor, 8, 89, 108, 111, 114, 120, 137, 139, 187, 191, 193, 207, 213

project stakeholders, 12, 139, 173, 249
project success, 26, 45–47, 53, 60, 79, 80, 87, 89, 100, 122, 128, 130, 171, 173, 175, 182–190, 216, 219, 228, 240, 277, 279, 288, 290, 293–295
project success assessment, 173, 175
project success criteria, 122, 183
project team (PT), 46, 56, 61, 67, 92, 93, 100, 114, 118, 119, 124–126, 129, 132, 135–139, 141, 144–147, 150, 155, 161, 162, 169, 173, 174, 183, 184, 188, 190, 198, 214, 227, 248, 249, 251, 262, 270
project type, 5, 13, 55–57, 75, 135, 163, 181, 196, 202, 231, 278, 295
provisional acceptance, 154, 161, 163–165, 169
PT. *See* project team

Q

quality, 7, 9, 18, 45, 48, 78, 101, 102, 125, 127, 128, 132, 138, 144–147, 152, 155, 158, 159, 162, 174, 184, 214, 215, 220, 238, 251, 253, 265

R

real projects in the real world, 26, 294
reconciliation, 103, 172–174, 176, 177, 231, 237, 244, 274, 290
Requirements Stage, 39, 61, 77, 123, 127, 129, 132, 153, 163, 202, 208, 214, 273, 283, 286, 289
resource management, 12, 15, 17, 97
resources, 5, 8, 9, 11, 18, 19, 22, 26, 48, 52–54, 63, 67, 71, 78, 86, 92, 97, 101–103, 109, 110, 113–115, 117, 120, 124, 129, 134, 135, 138, 139, 144, 148, 150, 151, 162, 165, 174, 181, 193, 194, 197, 201, 202, 212, 213, 219, 222, 228, 229, 238, 239, 249–251, 253, 259, 267, 288, 289
risk assessment, 101, 119, 245, 246
risk management, 13, 70, 76, 115, 150, 152, 174, 216, 223, 229, 245, 246, 248, 265, 273
risk management across the stages, 246

S

SAD. *See* stage authorization document
safety, 72, 88, 95, 97, 118, 128, 135, 138, 144, 146, 147, 152, 158, 165, 174, 213, 246, 247, 253
sample large-complex project, 209–211
sample project, 45, 272, 285, 286
SAVE. *See* Society for the Advancement of Value Engineering International
scope creep, 238, 240–242, 244
scope of work, 75, 78, 92, 102, 121, 125, 126, 132, 145, 146, 149, 215
scoring model, 262
Scrum, 5, 6, 8, 46, 61, 77, 134, 239, 297
SDP. *See* stage detailed plan
second dimension, 22, 40, 52, 58, 83, 85, 87, 90, 103, 177, 181, 182, 185, 186, 205
service provider, 34, 47, 61, 65, 102, 144, 149, 150, 154, 162, 165, 183, 184, 212, 222, 227, 257, 271–275, 282
Seven Elements of Project Management Maturity™, The, 53, 80, 86, 294, 296
shifting paradigm, 295
simulating CAMMP™, 285
simulating managing across the stages, 208
size and complexity, 5, 40, 81, 132, 262, 278
small-simple project, 40, 74, 75, 261, 264, 265
SMP. *See* stage management plan
social or community evaluation, 117
Society for the Advancement of Value Engineering International (SAVE), 8, 26, 52, 57, 195, 231
specialization in project management, 257
Stages
    Close, 37, 38, 61, 69, 79, 80, 99, 103, 171–173, 177, 186, 212, 216, 231, 244, 248, 274, 288, 290

Concept, 76, 107, 109
Definition, 77, 89, 124, 135, 137, 143, 151, 155, 160, 202, 215, 222, 223, 227, 237, 241, 242, 265, 273, 286, 288, 289
Feasibility, 76, 107, 113, 140, 145, 160, 207, 208, 213, 214, 231, 247, 248
Implementation, 78, 113, 130, 135–137, 144, 150, 153, 154, 159–161, 163, 165, 171, 174, 212, 215, 227, 272, 274, 287, 289
Initial Operations, 78, 163, 167, 215, 218, 279, 288
Operational Readiness, 78, 135, 163, 164, 167, 212, 215, 274, 279, 287, 290
Requirements, 39, 61, 77, 123, 127, 129, 132, 153, 163, 202, 208, 214, 273, 283, 286, 289
Strategy, 77, 89, 122, 127, 131, 215, 227, 265, 273, 289
stage authorization document (SAD), 95, 99, 100, 102, 114, 115, 131, 144, 154, 206, 208, 218, 234, 259
stage charter, 114, 115, 206–208, 213
stage control, 107, 233, 234
stage deliverable, 74–76, 79, 107, 113, 115, 131, 132, 154, 181, 218, 265, 285, 287
stage detailed plan (SDP), 92, 95, 99, 101–103, 114, 115, 124, 132, 144, 154, 173, 206, 208, 213, 214, 218, 226, 234
Stage Gate 1, 69, 80, 110, 111, 114, 121, 141, 157, 217, 221, 272
Stage Gate 2, 80, 114, 119, 120, 141, 158, 217, 219, 221, 222, 224, 226, 227, 247, 272, 273, 286
Stage Gate 3, 123, 124, 129, 130, 141, 217, 221, 273
Stage Gate 4, 131, 132, 140, 141, 144, 158, 217, 219, 221, 222, 224, 227, 273, 286
Stage Gate 5, 143, 144, 151, 152, 154, 218, 219, 221, 223, 224, 227, 240, 243, 273
Stage Gate 6, 153, 154, 161, 163, 165, 169, 218, 287
Stage Gate 7, 154, 161, 167, 169, 171, 218
Stage Gate 8, 154, 167, 169, 171, 218
Stage Gate 9, 99, 171, 173, 176, 186, 190, 218
Stage Gate Process, 68, 70, 218
stage gates, 6, 14, 58, 59, 67–70, 74–76, 79, 80, 88, 89, 95, 99, 100, 108, 110, 113–115, 120, 121, 123, 124, 129–132, 141, 143, 144, 151, 153, 154, 158, 161–163, 165, 169, 171–173, 176, 177, 181, 190, 208, 212, 214, 215, 217–223, 227, 228, 247, 250, 272, 273, 275, 278–280, 283, 286, 287
stage gates approvals, 217, 223
stage management plan (SMP), 36, 92, 95, 99, 101–103, 114, 115, 123, 124, 132, 144, 154, 206, 208, 213, 214, 218, 226, 234
stage scope, 6, 102, 145
stage versus project control, 233
stage versus project estimates, 225
stage versus project risks, 246
stakeholder alignment, 77, 123, 124, 130
Stakeholder Alignment Gate, 77, 123, 124
stakeholders, 12, 13, 15, 17, 18, 63, 65, 67, 69, 76, 77, 101, 102, 121–127, 129, 130, 137, 139, 141, 149, 161, 162, 173, 174, 217, 218, 242, 249, 250, 266, 280
stakeholders across the project stages, 249
stakeholders, appropriate, 126, 127, 129
standard, 3–6, 8, 11, 13, 15, 17, 19, 21, 22, 52, 54, 58, 59, 62, 66, 74, 75, 79–81, 86, 87, 91, 127, 133, 151, 153, 158, 171, 181, 189, 198, 208, 219, 221–223, 225, 226, 265, 272, 273, 277–283, 285, 288, 289
Standard for Sustainability in Project Management (P5), 8, 22, 66, 87, 198
standard model, 58, 59, 62, 74, 75, 79–81, 153, 208, 219, 222, 223, 265, 272, 277, 279, 280, 282, 283, 285, 288
strategic alignment, 60, 69, 80, 110, 114, 115, 289

strategic aspects in CAMMP™, 80
Strategy Stage, 77, 89, 122, 127, 131, 215, 227, 265, 273, 289
subcontractor, 275
subject areas, 6, 17, 18, 22, 32, 44, 88, 92
subprocess, 90, 100, 103
success factors, 64, 128, 279, 295
SUKAD, 5, 14, 22, 44, 47, 52, 54, 57, 64, 69, 72, 86, 87, 100, 109
SUKAD CAMMP™ Model, The, 57
support processes, 19, 25, 57
sustainability, 8, 22, 25, 52, 53, 57, 87, 90, 95, 97, 114, 118, 128, 134, 146, 150, 182, 197–199, 213, 214, 295
sustainability and CAMMP™, 198
sustainability considerations, 8, 118, 198, 214
sustainability strategy, 134

T

tailor CAMMP™, 277
tailoring, 5, 15, 40, 43, 44, 54, 58, 59, 79, 80, 223, 257, 264, 265, 272, 275, 277, 281, 282, 293
technical and functional personnel, 114
technology evaluation, 118
"think at two levels", 206

threats and opportunities, 245, 247
three-dimensional model, 22, 27, 40, 55, 57, 62, 73, 177, 205, 293
time-phased budget, 140, 149, 155, 156
traditional change management, 243
typical process within a stage, 95

U

universal methodological approach, 55, 73, 81

V

value-improving practices, 70, 128

W

Waterfall, 5, 30, 61, 125, 126, 258, 297
Waterfall and Agile, 297
Waterfall versus Agile, 258
WBS. *See* work breakdown structure
what is customizable and adaptable, 279
why change management is critical, 240
why customize and adapt the model, 278
work breakdown structure (WBS), 6, 31, 102, 126, 140, 146, 149, 150, 152, 155
writing and publishing a book project, 285